An Atlas of
SIGMOIDOSCOPY AND CYSTOSCOPY

THE ENCYCLOPEDIA OF VISUAL MEDICINE SERIES

An Atlas of
SIGMOIDOSCOPY AND CYSTOSCOPY

Brigitte E. Miller, MD
Wake Forest University Medical School,
Winston-Salem, NC, USA

SOUTH UNIVERSITY
709 MALL BLVD.
SAVANNAH, GA 31406

The Parthenon Publishing Group
International Publishers in Medicine, Science & Technology

A CRC PRESS COMPANY
BOCA RATON   LONDON   NEW YORK   WASHINGTON, D.C.

Library of Congress Cataloging-in-Publication Data
Miller, Brigitte E.
An atlas of sigmoidoscopy and cystoscopy / Brigitte E. Miller
   p. ; cm. -- (The encyclopedia of visual medicine series)
   Includes bibliographical references and index
   ISBN 1-85070-927-0 (alk. paper)
   1. Sigmoidoscopy--Atlases. 2. Cystoscopy--Atlases. I. Title.
   II. Series
   [DNLM: 1. Sigmoidoscopy--Atlases. 2. Colonic diseases--
diagnosis--Atlases. 3. Cystoscopy--Atlases. 4. Rectal diseases--
diagnosis--Atlases. 5. Urologic Diseases--diagnosis--Atlases
WI 17 M647a 2001]
   RC804.S47 M55 2001
   616.3´407545–dc21   00-068673

British Library Cataloguing in Publication Data
An atlas of sigmoidoscopy and cystoscopy. – (The encyclopedia
of visual medicine series)
   1. Sigmoidoscopy 2. Cystoscopy
   I. Miller, B. E.
   616.3´4´07545

   ISBN 1-85070-927-0

Published in the USA by
The Parthenon Publishing Group Inc.
One Blue Hill Plaza
PO Box 1564, Pearl River
New York 10965, USA

Published in the UK and Europe by
The Parthenon Publishing Group Limited
Casterton Hall, Carnforth
Lancs., LA6 2LA, UK

Copyright © 2002 The Parthenon Publishing Group

No part of this book may be reproduced in any form without permission from the publishers, except for the quotation of brief passages for the purposes of review.

Typeset by Siva Math Setters, Chennai, India
Printed and bound by T.G. Hostench S.A., Spain

Contents

List of principal contributors	7
Foreword *F.W. Ling*	9
Preface	11

Section I A Review of Sigmoidoscopy — 13

1	Indications for sigmoidoscopy *B.E. Miller*	15
2	Instruments for sigmoidoscopy *B.E. Miller*	25
3	Patient preparation for sigmoidoscopy *B.E. Miller*	33
4	The sigmoidoscopy examination *B.E. Miller*	41

Section II A Review of Cystoscopy — 51

5	Indications for cystoscopy *B.E. Miller*	53
6	Equipment for cystoscopy *G. Portera*	57
7	Patient preparation for cystoscopy *B.E. Miller*	61
8	The cystoscopy examination in the female *B.E. Miller and W.A. Katsanis*	65

Section III Sigmoidoscopy and Cystoscopy Illustrated — 73

R.F. Werkmann, R.R. Dmochowski, B.E. Miller, W.A. Katsanis and G. Portera

Index	100

List of principal contributors

Roger R. Dmochowski, MD
Fort Worth, TX
USA

Brigitte E. Miller, MD
Wake Forest University Medical School
Winston-Salem, NC
USA

Greg Portera, MD
Memphis, TN
USA

Robert Werkmann, MD
Milton S. Hershey Medical Center
Hershey, PA
USA

Foreword

The world of medicine continues to evolve at a dizzying pace. Despite the advent of the Internet, the explosion of new information and the changing relationships between physician and patient, there remains the need for effective dissemination of information in easily-digestible packages. Such is the case with this work by Brigitte E. Miller, and colleagues.

To put this text in proper perspective, the reader would do well to get to know Dr Miller, a gynecologic oncologist who is the embodiment of excellence – she is an outstanding clinician loved by her patients; she is an exemplary teacher revered by her residents and students; and she is a joy to have on the faculty of a medical school as she seeks to create new knowledge. While juggling all of these roles, her successes as wife and mother are additional notable accomplishments.

This book is an outgrowth of Dr Miller's unflagging energy to improve patient care. She has assembled information from diverse sources to help direct the education of those who would pursue expertise in sigmoidoscopy and cystoscopy. From the preoperative consents and other paperwork to the intraoperative findings, this book fills a significant void on our library shelves. The words and pictures complement each other to present the fundamentals of critically important procedures.

I am confident that this unique physician has successfully produced an equally unique resource for her fellow physicians. I congratulate her on this effort of love.

Frank W. Ling, MD
*UT Medical Group Professor and Chair
Department of Obstetrics and Gynecology
University of Tennessee College of Medicine
Memphis, TN, USA*

Preface

Physicians have previously concentrated on treating medical problems as they arose. Care was given in reaction to a problem or disease. However, many problems can be prevented now, or at least significantly ameliorated, by screening for precursor lesions and early intervention. A greater part of future medicine will be proactive, and preventive health care will become more and more important. For example, colon cancer is the third most frequent malignancy in men and women, detected in over 130 000 patients every year in the USA. Although a positive family history identifies men and women at increased risk, most cases are diagnosed in patients without such a history. Colon polyps are another important risk factor and, during endoscopy, polyps can be removed, leading to an overall decreased risk of developing cancer. The diagnosis of these early changes in today's environment becomes even more important, as there is emerging evidence that medical treatment may be effective in further reducing the risk. This information is one more reason to closely follow screening recommendations.

This book is not written for the specialist and does not imply that after reading a few pages the specialist's knowledge and skills can be attained. Nevertheless, students and residents involved in sigmoidoscopy and cystoscopy will benefit from having key information available within one resource. Another of the anticipated audiences for this book is the primary care provider, who is in a pivotal position to evaluate the risk status of the patient and recommend screening as necessary. Patients listen best to the recommendations of their doctor! In addition, access to a specialist can sometimes be difficult. We hope that this book will be helpful in allowing nonspecialists to gain some initial knowledge about sigmoidoscopy and cystoscopy, and organize an examination area in the office. Furthermore, we hope that the figures will be helpful to learn more about the appearance of the sigmoid colon and the bladder, as well as the most important pathologic findings. Needless to say, no picture can replace experience, and no book can replace good training and constant practice. This book should not replace the specialist, but it should help to make appropriate referrals and give more patients access to screening exams.

Modern medicine is best practiced in a team environment. Therefore this book is also a team effort. I would like to thank all the contributors in academic medicine and private practice, who helped so much. We are all grateful to our editor, who helped in so many ways, and to Parthenon Publishing, who gave us the opportunity to compile our experiences.

Brigitte E. Miller
Wake Forest University Medical School
Winston-Salem, NC, USA

Section I A Review of Sigmoidoscopy

CHAPTER I

Indications for sigmoidoscopy

Brigitte E. Miller

Screening examinations account for the majority of sigmoidoscopies performed by primary care physicians. The literature regarding colon cancer prevention and early detection, including the role of flexible sigmoidoscopy, is reviewed in the following chapter. Additional indications for flexible sigmoidoscopy, are discussed and compared with other means of evaluation in an effort to describe the most efficacious way to establish the diagnosis of colorectal disease in average and high-risk patients. As with any invasive procedure, there are contraindications, absolute and relative, which are discussed in the last section. However, no advice can replace good clinical judgement and one should never hesitate to ask for help.

SCREENING

In the following chapter, the literature regarding colon cancer screening is reviewed, and information necessary for patient education and motivation is summarized. The influence of the primary care physician is critical to increase patient compliance in any screening program.

Screening programs are indicated for diseases that are serious; have a long preclinical phase; are treatable, with much better results if detected in an early stage; and occur frequently enough in the general population to merit screening. Furthermore, for flexible sigmoidoscopy to be an adequate test, it should give valid, reliable, reproducible results; should be easy to administer, easy to tolerate and inexpensive.

Serious disease

Colorectal cancer is a very serious disease. Calculations by the National Cancer Institute's Surveillance, Epidemiology, and End Result Program[1] (SEER) projected that, in the year 2000, 56 300 patients would die from colorectal cancer. Colorectal cancer accounts for 10% of all cancer diagnoses and deaths.

Long preclinical phase

There is suggestive evidence that many colon cancers develop from adenomas to invasive carcinoma over a period of 5–10 years[2]. Using molecular biological techniques, Vogelstein and colleagues[3] identified a sequence of events leading from normal mucosa to adenoma and finally to invasive cancer. The National Polyp Study[4] as well as other studies have shown that the removal of colonic adenomas followed by surveillance colonoscopy reduces the incidence of colon cancer. A report from the Mayo Clinic[5] revealed that, among 220 patients in whom colon adenomas were followed conservatively, colon cancer developed in 2.5%, 8% and 24% after 5, 10 and 20 years, respectively. However, in patients who underwent polypectomy, a significant reduction in the incidence of colon cancer was noted: only five asymptomatic early stage lesions were discovered in 1418 patients after a median follow-up of 6 years[4]. Overall, about 10% of adenomas will give rise to carcinomas over a 15-year time span[6]. Large adenomas (above 1 cm), villous adenomas and those with dysplasia carry an increased risk of malignancy by 3%, 17% and 37%[7]. Hyperplastic, lymphoid, or inflammatory polyps do not increase the risk of colon cancer.

The fact that there are remnants consistent with adenomatous changes in most early colon cancers is regarded as further evidence that most colon cancers develop from adenomas. Bedenne and co-workers[8] examined over 1600 colon cancers and found adenomatous remnants in most of the early exophytic tumors.

Results of treatment

Early-stage colon cancer can be treated with good results. Cancer statistics show that the overall survival rate for colon cancer is 62%; however, for localized disease confined to the colon, a 90% survival rate can be expected[1].

Colon cancer prevalence

The lifetime risk of colon cancer is about 1 in 7[1]. It was estimated that 130 000 patients would be diagnosed with a colorectal malignancy in 2000. Screening sigmoidoscopy detected adenomas in 5.4% of over 4000 asymptomatic subjects aged 50 to 65 and malignancies in 0.32%[9].

Validity of results

Sigmoidoscopy is very accurate for the portion of colon examined. Lesions of > 5 mm in size are rarely missed, and this holds true if the examination is performed by primary care physicians. Hawes and associates[10] trained residents in sigmoidoscopy and noted that competence was achieved after 24–30 examinations. At that point, residents detected 93–100% of all lesions as described by an experienced endoscopist. With biopsy of all lesions, the examination is also extremely specific. In addition, the endoscopic evaluation is not only useful in the early diagnosis of colon cancer, but also an important preventive measure, as polyps and adenomas can be detected and removed, thus decreasing the risk of cancer in the future.

However, between 20 and 50% of adenomatous lesions are seen in the proximal colon only, and obviously will escape detection by sigmoidoscopy[11]. Proximal malignant lesions out of reach of the sigmoidoscope increase with age from 39% at age 65 to 50% at age 85[12]. In addition, there is evidence for an overall increasing incidence of proximal lesions over the past several decades[13]. Therefore, sigmoidoscopy can reduce the colon cancer rate by only 45% at the maximum[14]. Fecal occult blood testing (FOBT) has been shown to be effective for colon cancer screening in several randomized studies, reducing mortality by 33%[15]. Although FOBT leads to an earlier diagnosis, it has no preventive effect. FOBT is not sensitive for the diagnosis of adenomas because these do not bleed consistently. Therefore, overall detection rates are not as high as with sigmoidoscopy[16]. Combining FOBT with screening sigmoidoscopy has been shown to increase the detection rate of more proximal lesions and to increase the overall detection rate[17].

Ease of administration

Although it is more involved than FOBT, sigmoidoscopy is an uncomplicated examination. Preparation for sigmoidoscopy with an enema is usually well tolerated and easily completed. Because sigmoidoscopy requires no extensive preparation (necessary for colonoscopy) and no diet modification or follow-up testing, the procedure can be easily included in a routine check-up visit. Overall, flexible sigmoidoscopy takes between 4 and 10 min and, in the vast majority of patients, no premedication is necessary. Although an adequate office set-up for sigmoidoscopy is necessary, equipment costs are moderate.

Toleration of screening

Of course, a screening test is only as good as compliance. Compliance rates for FOBT ranged between 80 and 92% in several studies. There are fewer data available for sigmoidoscopy; compliance rates ranged between 12 and 69%[18]. A better compliance was noted among patients seeking regular medical care, patients who received information and reminders from their physician, or patients in work site-based studies. Reporting on twin siblings of colon cancer patients, Richardson and colleagues[19] noted that only 69% had ever had a sigmoidoscopy and only 43% were compliant with yearly FOBT. This was quite surprising, as the participants in this study had been aware of the discomfort associated with colon cancer or had even experienced the death of their sibling due to metastatic malignancy.

A physician's attitude plays a very important role. Triezenberg and co-workers[20] evaluated cancer screening practices of family physicians and, not surprisingly, noted that those with the most aggressive approach detected a larger number of asymptomatic cases, as well as a greater overall number of invasive malignancies. The only way to improve these numbers is by patient education, especially through the efforts of all primary care providers.

Cost of screening

Especially in our current medical environment, the cost of screening is important. Evaluation for fecal occult blood is the most cost-effective screening method. The addition of flexible sigmoidoscopy has not only increased the cancer prevention rate by 30%, but also increased the cost per death prevented by 15%[21]. Overall costs for screening and follow-up examinations were calculated to be $20.00 per patient per year for those with FOBT and $48.00 per patient per year for those with both FOBT and sigmoidoscopy[22]. In comparison, yearly mammography costs about $150.00 and a yearly gynecological examination with Pap smear about $85.00. In 1989, Gupta and co-workers[23] estimated that it would cost $47 000.00 to detect a potentially curable colon cancer using screening sigmoidoscopy. This is in the same range as the benefit expected from yearly mammography. Also, these figures are significantly affected by patient compliance, especially with the FOBT, because it has to be done every year. Here again, patient education and the recommendations given by the primary care physician are essential.

Overview of screening

Although no prospective randomized studies are available, there have been several large non-randomized studies pointing to the probable benefit of screening sigmoidoscopy. Unfortunately, these studies are limited by a self-selection bias because patients with very mild symptoms may ask to be screened, thus increasing the apparent benefit of screening. In the largest study, reported by Gilbertsen and Nelms[24] on over 100 000 examinations in 20 000 patients, none of the patients undergoing sigmoidoscopy and treatment for adenomas died of colorectal cancer. Another early study done in 1960 by Hertz and colleagues[25], who screened 26 000 patients with rigid sigmoidoscopy, detected 58 cancers, i.e. one malignancy in 450 examinations. All cases underwent complete surgical resection and the overall 5-year survival rate was 88%. In a third study performed by the Kaiser Permanente Medical Care Program[26], a study group was offered sigmoidoscopy and the control group was not. After 16 years, there was a significant decrease in the number of deaths due to colorectal cancer within the screened group, and malignancies were diagnosed at an earlier stage. Unfortunately, all these studies were not randomized and were impaired by low patient compliance rates.

Observational studies may overestimate the benefit of sigmoidoscopy. In small case–control studies, Newcomb and colleagues[27] and Selby and colleagues[28] noted a reduction in colon cancer mortality by 80 and 70%, respectively, after following patients for 10 or more years. Another very large case–control study evaluated over 32 000 patients at a veterans' medical center[29] and noted that there was a significant difference between the frequency of endoscopic procedures in patients diagnosed with colon cancer when compared to the control group. It was estimated that endoscopy reduced the risk of colorectal cancer by 50%, an effect lasting for about 6 years. Again, all these retrospective studies are subject to the risk of selection bias and incomplete data collection, but are evidence of the achievements of common medical practice.

Length of scope used

Initially, rigid sigmoidoscopy was used for screening. Compliance was poor, however, owing to significant discomfort, and results were suboptimal, owing to the short length of bowel evaluated.

Two to three times more lesions can be detected when a flexible sigmoidoscope is used of up to 60 cm in comparison to a rigid sigmoidoscope of up to 25 cm. Sarles and associates[30] compared flexible sigmoidoscopy with a 30-cm instrument and a 60-cm instrument. Although the longer scope was more difficult to handle, there were no significant differences regarding training time, cost, or complication rates. The examination that used the 60-cm instrument took about 4 min longer. More importantly, a greater number of polyps was detected with the 60-cm instrument. Therefore, a 60-cm flexible sigmoidoscope is recommended.

Who should be screened?

At this point, everyone should be screened, except patients with other life-threatening diseases who would gain little survival benefit from early detection of colon cancer. General recommendations are summarized at the end of this chapter.

When should screening start?

The incidence of colon cancer and adenomas rises with age, especially during the sixth decade of life when adenomas are seen in over 10% of patients. Ransohoff and Lang[31] calculated that between the ages of 40 and 49, 8000 sigmoidoscopies would have to be done to prevent one death from colon cancer. This figure drops to 2000 between the ages of 50 and 59, and to as low as 1000 between the ages of 60 and 69. Therefore, it seems reasonable to initiate screening at age 50. The National Polyp Study revealed a significantly higher incidence of dysplasia in patients over the age of 60[32]. Using literature data for statistical evaluation of screening efficacy, Eddy[33] likewise concluded that there was little benefit to initiating screening before the age of 50 in an average-risk patient. An upper age limit for screening has not been defined and should depend on the individual performance status of the patient and presence of other significant health problems.

Atkin and co-workers[34] calculated the benefit of a single sigmoidoscopy between the ages of 55 and 60, as the incidence of colon polyps increases significantly in the sixth decade. All patients with large (> 1 cm) or high-risk polyps (villous histomorphology) underwent complete colonoscopy and further surveillance. With this regimen, it was estimated that about 5500 colorectal cancers could be prevented every year in the UK at a cost of $5500.00 per cancer prevented. By evaluating 621 asymptomatic patients with negative hemoccult tests, Rex and associates[35] noted cancer in 0.6% and adenomas in 25%. Fifty per cent of the lesions were noted proximally in the sigmoid. These authors therefore recommended a single colonoscopy between the ages of 55 and 60 with follow-up colonoscopic examinations for patients with polyps. Again, no randomized studies have been done to confirm these hypotheses.

Screening intervals for sigmoidoscopy

Selby and Friedman[26] recommended a screening interval of 3–5 years, but also mentioned that an interval of 10 years may be as effective. In view of the slow progression from polyps to cancer, Lieberman[14] also suggested extending the screening interval to 5 or even 10 years. At this point, there is not sufficient information available to make a definite statement about the effect of a 10-year screening interval. Most studies were done using examinations every 5 years, and this remains the recommendation at present.

Family history of colon cancer

In about 20% of all patients with colon cancer, genetic factors may play a role. Wu and co-workers[36] noted an increased prevalence of polyps, as well as a higher rate of proximal lesions, in first-degree relatives of colon cancer patients. Few abnormalities were noted in patients under the age of 40. Therefore, they recommended colonoscopy for all first-degree relatives above the age of 40. Rozen and associates[37] agreed with these recommendations. They found an overall three-fold increased risk for colorectal adenomas and malignancy in family members of cancer patients. However, they detected very few polyps and no malignancies in patients under the age of 40. In contrast, Rex and co-workers[35] noted no significantly increased risk of polyps or malignancy in patients with a single, first-degree relative with colon cancer, except when the malignancy occurred before the age of 60.

True familial syndromes account for a small percentage of colon cancers, but convey a much higher risk; therefore, different guidelines apply. The diagnosis can be made only after careful evaluation of family history regarding all malignancies. This should be part of every screening examination. The hereditary site-specific colon cancer (hereditary non-polypotic colon cancer, HNPCC, Lynch I syndrome) has to be distinguished from the cancer family syndrome (Lynch II), which also conveys a higher risk of other adenocarcinomas, including breast and endometrial adenocarcinomas. Inherited syndromes should be strongly suspected if there are two or more affected first-degree relatives or if colon cancer is diagnosed before the age of 30. In the future, genetic screening tests may become available to identify these high-risk groups. Adequate counseling is very important and, therefore, these patients are best followed by a specialist. As malignancy occurs at a much younger age in these syndromes, screening may have to be initiated as early as age 20.

Patients at increased risk

As discussed above, the presence of an adenoma increases the risk of colon cancer. If an adenoma is noted on sigmoidoscopy, the entire colon should be evaluated with colonoscopy. Even small lesions can be reliably detected and diagnosed, and treated with excisional polypectomy. A double contrast barium enema is not as sensitive, especially for smaller lesions, but may become necessary when a complete colonoscopy is not possible or it failed, for technical reasons.

After treatment for colon cancer, patients are at risk of recurrence, approaching 30% after curative resection. Fifty per cent of the recurrences will occur during the first post-treatment year and another 20% during the next year. Recurrences after 5 years are rare, but, in about 5% of the patients, a metachronous tumor will develop an average of 11 years after the initial diagnosis[38]. A careful history and general physical examination should be carried out, and FOBT and carcinoembryonic antigen (CEA) level should be assessed every 2–4 months during the first 2 years after treatment, as well as a full colonoscopy performed 1 and 2 years after treatment, and then at 3-year intervals[39].

A long history of idiopathic inflammatory bowel disease increases the risk of colon cancer. This holds true especially for chronic idiopathic ulcerative colitis with a universal distribution (pancolitis) where the risk rises after 10 years and increases up to 12% after 25 years. Dysplastic areas are difficult to detect and may be missed even if sequential biopsies at short intervals (10 cm) are taken throughout the entire colon[40]. It is recommended that these patients be followed by physicians with special training. Crohn's disease also increases the risk of malignant degeneration[41].

Women with a history of gynecological cancer or breast cancer[42] and those treated with pelvic radiation therapy may be at a slightly increased risk of colon cancer. It is not known whether more frequent screening is necessary, but the current recommended intervals should be strictly followed.

Who should do the screening?

With the number of examinations necessary, most of these will have to be carried out by the primary care physician. Schapira and co-workers[43] found a high compliance rate when primary care physicians recommended cancer screening. Unfortunately, the recommendations for sigmoidoscopy were not very strict.

Personal contact by the physician and continued follow-up also should increase the compliance rate.

Conclusion

Flexible sigmoidoscopy is valuable in the early diagnosis and prevention of colon cancer. Although no randomized study is available to definitely prove its effect, the evidence from well-conducted case–control studies is compelling. FOBT has reduced colon cancer mortality, but has no effect on prevention. Combining sigmoidoscopy and FOBT seems to give the best results, although cost effectiveness is still under discussion. Preliminary data from a controlled European trial comparing FOBT with and without sigmoidoscopy revealed a significantly higher diagnostic yield with the addition of sigmoidoscopy in spite of compliance rates of only 19 to 41%[44]. Using sigmoidoscopy and FOBT in over 21 000 patients, Winawer and co-workers[17] were able significantly to increase the likelihood of a cancer diagnosis at an early stage. In the future, a single colonoscopy with follow-up of high-risk patients may develop into another valid alternative. Ten million examinations would be necessary per year if all persons between the ages of 50 and 75 were to be screened every 5 years in the USA[32]. As the experience in the Kaiser Permanente system revealed, screening may be worthwhile in a managed care setting. Some of the major insurance groups also pay for screening examinations. Recent Federal Medicare legislation (effective 1998) allows physician and hospital reimbursement for the 'screening' of average-risk patients (beginning as early as 50 years of age) with flexible sigmoidoscopy and of 'high-risk' patients with colonoscopy or flexible sigmoidoscopy plus barium enema.

Among experts, there is still heated discussion as to the best screening system for colon cancer. The National Cancer Institute has launched a large screening trial evaluating sigmoidoscopy (PLCO trial), but we will have to wait several years for the results. The newest recommendations of the American College of Gastroenterology and the American Cancer Society[45] are as follows:

Average risk

(1) FOBT is recommended every year in patients over the age of 50, and/or:
(2) Flexible sigmoidoscopy is recommended for all patients aged 50 or over, every 5 years. If other options are chosen, double contrast barium enema should be repeated every 5–10 years, colonoscopy every 10 years.
(3) Digital rectal examination should always be performed during endoscopy or barium enema.

Moderate risk

Patients with a history of polyps or colon cancer should have the entire colon evaluated with colonoscopy. After removal of a small polyp, the examination should be repeated after 3 years and, when negative at that time, the examinations should return to regular screening intervals for low-risk patients. If the polyp was larger than 1 cm or if multiple lesions were noted, screening should be carried out 1 and 3 years after removal and then every 5 years. The same regimen should be followed after curative resection of colon cancer. If this is not possible, then a double contrast barium enema and sigmoidoscopy should be ordered.

Patients with a first-degree relative with colon cancer or polyps who are below the age of 60 or with two or more affected first-degree relatives should have the entire colon evaluated beginning at age 40 or at least 10 years before the youngest case in the family, whichever is earlier. Follow-up examinations should be performed every 5 years. Patients with other relatives affected by colon cancer should be screened according to average risk recommendations; however, screening may be considered earlier.

High risk

Patients with pancolitis should be evaluated by colonoscopy every 1–2 years beginning 8 years after the diagnosis of inflammatory bowel disease. If there is only left-sided colitis, then evaluation should start about 12–15 years after the diagnosis.

Patients with familial polyposis should be evaluated at puberty and followed with colonoscopy every 1–2 years. Colectomy may have to be considered.

Patients from a family with hereditary non-polyposis colon cancer should be counselled regarding genetic testing and be followed every 2 years from the ages of 21 to 40, and then yearly.

Updated guidelines as well as patient and reimbursement information are published on the Internet by the American College of Gastroenterology.

RECTAL BLEEDING

The differential diagnosis of bright red rectal bleeding (hematochezia) is extensive (Table 1)[46]. It is also important to consider an upper gastrointestinal source of bleeding with rapid gastrointestinal transit, especially in an unstable patient. In 3 to 5% of the cases, the small bowel is the origin of bleeding. Most often, however,

Table 1 Causes of rectal bleeding

Anal canal
    hemorrhoids
    fissures
    tumors
Rectum and colon
    diverticulosis/diverticulitis
    vascular malformations
    colitis
        idiopathic inflammatory bowel disease
        infectious
        antibiotic-induced
        ischemic
        radiation-induced
    tumors
        polyps
        malignancy
Diseases of small intestine or upper gastrointestinal tract

bleeding comes from the colon and anorectum. The differential diagnosis also depends on the patient's age. In infants, children, adolescents and young adults, Meckel diverticulum, inflammatory bowel disease, juvenile polyps, or disorders of the anal canal (anal fissure, internal hemorrhoids) are the most frequent causes of bright red rectal bleeding. In adults up to the age of 60, rectal bleeding is caused mainly by diverticulosis, inflammatory bowel disease, polyps, carcinoma, arteriovenous malformations and especially by fissures or hemorrhoids. In patients over the age of 60, hemorrhoids, diverticulosis, angiodysplasia, carcinoma and ischemia are the most frequent diagnoses. Among 2200 patients aged 35 to 75 examined with colonoscopy, Shinya and colleagues[47] noted polyps in 33%, hemorrhoids or anal pathology in 27%, carcinoma in 19% and inflammatory bowel disease and proctitis in 13%. Richter and colleagues[48] found diverticulosis (47%), angiodysplasia (12%) and adenomas (8%) among the most frequent causes of lower gastrointestinal bleeding. In patients over 80 years old, Bat[49] found a malignancy in 28.7% of the cases.

Evaluation of patients with rectal bleeding should always include a complete history regarding bowel habits and the duration and type of bleeding. It should be noted whether the bleeding is related to bowel movements or associated with tenesmus or abdominal pain or cramping. Fresh blood on top of the stool is usually related to bleeding from the rectum or sigmoid. It is mainly caused by hemorrhoids, but can also be seen with fissures, infectious problems, radiation-induced changes, polyps and malignancies. This should be followed by a careful general physical examination to exclude infectious lesions or an acute abdomen. With a rectal examination, the gastrointestinal origin of the bleeding should be confirmed. The hemoglobin level should be checked and coagulation parameters should be evaluated in the presence of severe bleeding.

If anoscopy combined with flexible sigmoidoscopy reveals a definite area of bleeding and no evidence of polyps in a patient below the age of 50, then no further evaluation is necessary, as the risk of colon cancer in these patients is low. It is important to include a retroflexed view of the anus at the time of sigmoidoscopy to ensure that small lesions close to the anus are not overlooked. However, in older patients, a complete evaluation by colonoscopy should be considered. This should also be done if rectal bleeding persists after adequate treatment, because in up to 30% of the patients an additional lesion may be noted in the colon[50]. If polyps are noted, especially those larger than 1 cm, a complete colonoscopy is indicated in all patients as the entire colon is always at risk. A barium enema is less useful in the diagnosis of rectal bleeding, as vascular malformations cannot be seen, bleeding from a diverticular lesion cannot be confirmed and often small polyps are overlooked. Significant lesions (i.e. polyps greater than 1 cm) were seen in 35%[51] of patients with rectal bleeding and a normal barium enema.

Subclinical bleeding leading to a positive hemoccult test or chronic anemia

The entire colon should be evaluated to assess for diverticulosis, polyps, arteriovenous malformations, chronic infections and also malignancy. In a Swedish study[52] using flexible sigmoidoscopy and double contrast barium enema, a sensitivity of 94% and specificity of 99% were noted in 530 patients. Carcinoma was diagnosed in 26 patients and an adenoma over 1 cm in diameter was diagnosed in 71 patients. Overall, three carcinomas as well as three adenomas were overlooked. Kewenter and colleagues[53] missed only 2.4% of the 83 malignancies detected among 1831 patients with a positive hemoccult test using the same technique. However, a significant percentage of patients undergoing barium enema will need colonoscopy for complete evaluation. Therefore, an initial colonoscopy is the more cost-effective examination, and remains the initial diagnostic method of choice for subclinical rectal bleeding.

Severe rectal bleeding

Patients with severe bleeding leading to hypotension, or older, frail patients with significant anemia or other medical problems, or those requiring transfusion should be referred to a specialist for consideration of emergency colonoscopy. Fluid resuscitation and rapid bowel preparation by large-volume oral–gut lavage may be necessary in this setting. This allows for prompt diagnosis as well as endoscopic treatment if necessary.

Rectal bleeding after pelvic radiation

Radiation therapy to the pelvis can lead to injury of the colon and rectum. Clinical signs such as rectal bleeding, diarrhea, pain and tenesmus usually occur 6–24 months after completion of therapy, but can develop many years later. After radiation for cervical cancer, the upper rectum and lower sigmoid are particularly at risk, owing to intracavitary radiation placed in the cervix and upper vagina. Pelvic radiation, as given for endometrial or prostate cancer, can also affect the lower ascending colon, descending colon, entire sigmoid and especially the terminal ileum. Although characteristic changes of radiation injury are often seen with sigmoidoscopy, an initial evaluation of the entire colon with endoscopy or barium enema should be performed to exclude other lesions. Biopsies are often necessary, but should be taken carefully, as delayed healing, increased ulceration and, rarely, fistula may develop, owing to the obliterative endarteritis of the irradiated tissue. Most of these patients should be referred to a specialist, as these examinations are often more difficult and painful, and sometimes endoscopic therapy is possible.

DIVERTICULAR DISEASE

Diverticular disease is usually confined to the lower left colon, where sigmoidoscopy can be employed to establish the diagnosis. Because diverticula are found in 33 to 50% of all persons over the age of 50, not all symptoms are necessarily due to this finding.

Acute diverticulitis is often caused by a small perforation of a single diverticulum. An intense inflammatory reaction follows to contain the inflammation, leading to significant scarring. Endoscopy during this stage has an increased risk of enlarging the perforation. Therefore, it is important to rule out any peritoneal signs by careful abdominal examination prior to the examination. Chronic diverticulitis often causes extensive scarring and fixation of the sigmoid, making it difficult to advance the endoscope and increasing the risk of inadvertent perforation. The barium enema is a good addition, as it gives an accurate picture of the length, redundancy and dilatation of the colon, as well as colon muscle spasms associated with the pathogenesis of diverticular disease[54]. On the other hand, all stenoses seen on barium enema in these patients should be further evaluated endoscopically regarding their clinical significance. Large diverticula can also render it difficult to identify the bowel lumen. All these facts can make sigmoidoscopy for diverticular disease challenging.

COLITIS

Chronic idiopathic inflammatory bowel diseases

Sigmoidoscopy with directed biopsy and a barium enema are recommended by the American Society for Gastrointestinal Endoscopy for the diagnosis of ulcerative proctocolitis or Crohn's disease of the colon. However, a complete colonoscopy is usually necessary, except in the acute stages of severe disease when the risk of perforation is high. Although ulcerative colitis almost always involves the rectosigmoid, these areas can be spared in Crohn's disease. Colonoscopy should be considered in all patients with an unexplained clinical course or with an abnormal or equivocal barium enema. Here, referral to a specialist is indicated, as these examinations are often more difficult to perform and the lesions more difficult to judge.

Infectious colitis

Acute infectious diarrhea is a very common complaint seen in general medical practice and, owing to the short, self-limited course, further evaluation is often not necessary. Severe infections, however, may lead to persistent bloody diarrhea, tenesmus and abdominal pain. Further evaluation is necessary in this situation. Clinical evaluation and cultures are often adequate to establish the diagnosis. Endoscopic findings are usually uncharacteristic and best seen in the proximal colon. Referral to a specialist should be considered.

Ischemic colitis

Sigmoidoscopy can be helpful in the diagnosis and clinical management of ischemic colitis, because the characteristic changes are noticed within the reach of the flexible sigmoidoscope in 80% of cases. In the presence of significant abdominal pain and signs of a systemic reaction, such as fever or leukocytosis, a surgical evaluation is also necessary.

Pseudomembranous colitis

In patients complaining of watery diarrhea after antibiotic therapy, infection with *Clostridium difficile* should be evaluated by culture and toxin determination. In a significant percentage of patients, characteristic lesions can be seen in the sigmoid and descending colon confirming the diagnosis.

Proctitis

Proctitis can be related to a variety of etiologies. Parasite studies and biopsies may be helpful to establish the diagnosis. In younger patients, Crohn's disease and ulcerative colitis always have to be considered; in older patients malignancy has to be excluded.

Venereal infections such as gonorrhea, syphilis, lymphogranuloma venereum, herpes and condyloma often involve the perianal area, in which case at least an anoscopy should be performed for complete evaluation in addition to appropriate cultures.

Patients with HIV infection often have diarrhea due to enteropathy; however, in patients with symptoms of colitis and rectal bleeding, sigmoidoscopy is indicated. Cultures and biopsies may be necessary to detect rare infections[55].

OTHER INDICATIONS

Abnormal barium enema

Some abnormal barium enemas require further work-up. If a lesion is identified in the lower descending colon, sigmoid, or rectum, sigmoidoscopy is ideal. However, the practitioner should be well versed in the technique and ready to take biopsies if necessary. In case of a stricture, a smaller colonoscope (pediatric colonoscope) may be necessary for complete evaluation of the area. Especially in the presence of deformation or narrowing, referral should be considered, as the area may be very difficult to negotiate, thus increasing the risk of complications.

Pretreatment evaluation for gynecological malignancies

Advanced gynecological malignancies sometimes involve the bowel. In patients with large vaginal or cervical malignancies involving the upper vagina and parametria, sigmoidoscopy is recommended to rule out rectal involvement. Rates of abnormalities are low, however, when the high rectal examination is completely normal. In addition to ruling out rectal spread of disease, sigmoidoscopy can detect conditions that may increase the risk of radiation complications. Finding diverticular disease or significant inflammation has to be taken into consideration when treatment is planned. In patients with endometrial or ovarian carcinomas, sigmoidoscopy is not necessary; however, patients should be current regarding their screening examinations. In patients with large bowel-related symptoms, colonoscopy or a barium enema is recommended.

Saltzman and colleagues reported on 212 patients with gynecological malignancies who underwent pretreatment colonoscopy. Abnormalities were noted in 11%; there were 17 polyps, five cases of colon cancer and two cases of cancer invading the colon. These authors recommended complete evaluation especially for patients over the age of 70 years[56].

Fistula

Anorectal fistulas can arise after long-standing, chronic infection of the glandular epithelium of the anal canal. Rectovaginal fistulas rarely develop after delivery or episiotomy. Prior to planning surgical repair, it is important to rule out chronic colonic inflammation, especially Crohn's disease. Therefore, sigmoidoscopy should be performed in every case and colonoscopy if Crohn's colitis is suspected.

Evaluation of symptomatic patients without evidence of bleeding

Rex and colleagues[57] evaluated 149 patients above the age of 40 who complained of constipation, diarrhea, abdominal pain, weight loss, or a combination of these signs and symptoms without evidence of clinical or subclinical rectal bleeding by either flexible sigmoidoscopy with barium enema or colonoscopy. Initial sigmoidoscopy detected more patients with diverticulosis than did colonoscopy. On the other hand, more adenomas were detected with colonoscopy. In patients who initially underwent flexible sigmoidoscopy and barium enema, an alternative procedure was necessary in 24%, whereas patients who underwent initial colonoscopy needed barium enema only in 6% of the cases. At this time, it is recommended that initial colonoscopy is the preferred and most cost-effective method of diagnosis in these patients, especially when a patient is over the age of 55, and the risks of a colonic malignancy are significantly increased. However, patients under the age of 50 without evidence of bleeding could safely be evaluated with sigmoidoscopy and barium enema. Overall, there was a very low prevalence of malignancy in this patient group (0.67%)[57].

CONTRAINDICATIONS

Clinical signs of colon perforation are an absolute contraindication to gas insufflation and sigmoidoscopy. Free abdominal air should be ruled out by X-ray in patients with suspicious findings on physical examination. Contraindications to colon preparation with enemas include suspicion of perforation, severe ulcerative colitis, toxic colonic dilatation or acute Crohn's disease, and acute diverticulitis with peritoneal signs. The examination should be carried out very carefully in patients with a possible obstruction. Sometimes the air or gas used for distension of the bowel can be trapped behind the partially obstructed area, leading to dangerous dilatation of the colon. In all these instances, referral to a specialist should be strongly considered, as these examinations are more difficult, and special instruments are often needed.

Some patients are less able to tolerate sigmoidoscopy, especially unstable patients after myocardial infarction or with severe respiratory compromise. Rarely, a patient is too uncooperative to undergo the examination safely. Rarely are coagulation defects so severe that sigmoidoscopy is contraindicated. Most patients on routine anticoagulation tolerate the procedure well, although a slightly higher risk of bleeding can be expected. Aspirin treatment should be discontinued 1 week prior to the examination if possible.

CONCLUSION

The main indication for flexible sigmoidoscopy in the primary care setting is screening for colorectal cancer. The primary care physician is in the optimal position to advise patients and ensure that the screening guidelines are followed.

Other indications for sigmoidoscopy include the initial evaluation of bright rectal bleeding in young patients and the diagnosis of inflammatory bowel disease or proctitis. Although sigmoidoscopy can be helpful in many other situations, often complete colonoscopy is necessary and referral should not be delayed.

SUGGESTED READING

Reintgen DS, Clark RA, eds. Cancer Screening. New York, NY: Mosby, 1996

Schapiro M, Lehman GA, eds. Flexible Sigmoidoscopy. Baltimore: Williams and Wilkins, 1990

REFERENCES

1. Greenlee RT, Murray T, Bolden S, Wingo PA. Cancer Statistics, 2000. *CA Cancer J Clin* 2000;50:7–33
2. Bronner MP, Haggitt RC. The polyp-cancer sequence: do all cancers arise from benign adenomas? *Gastrointest Endosc Clin N Am* 1993;3:611–22
3. Vogelstein B, Fearon ER, Hamilton SR, et al. Genetic alterations during colorectal-tumor development. *N Engl J Med* 1988;319:525–32
4. Winawer SJ, Zauber AG, Ho MN, et al., and The National Polyp Study Workgroup. Prevention of colorectal cancer by colonoscopic polypectomy. *N Engl J Med* 1993;329:1977–81
5. Stryker SJ, Wolff BG, Culp CE, Libbe SD, Ilstrup DM, MacCarty RL. Natural history of untreated colonic polyps. *Gastroenterology* 1987;93:1009–13
6. Morson BC. The evolution of colorectal carcinoma. *Clin Radiol* 1984;35:425–31
7. Eide TJ. Risk of colorectal cancer in adenoma-bearing individuals within a defined population. *Int J Cancer* 1986;38:173–6
8. Bedenne L, Faivre J, Boutron MC, Piard F, Cauvin JM, Hillon P. Adenoma–carcinoma sequence of 'de novo' carcinogenesis? *Cancer* 1992;69:883–8
9. Wherry DC, Thomas WM. The yield of flexible fiberoptic sigmoidoscopy in the detection of asymptomatic colorectal neoplasia. *Surg Endosc* 1994;8:393–5
10. Hawes R, Lehman GA, Hast J, et al. Training resident physicians in fiberoptic sigmoidoscopy. *Am J Med* 1986;80:465–70
11. Rex DK, Lehman GA, Hawes RH, Ulbright TM, Smith JJ. Screening colonoscopy in asymptomatic average-risk persons with negative fecal occult blood tests. *Gastroenterology* 1991;100:64–7
12. Cooper GS, Yuan Z, Landefeld CS, Johanson JF, Rimm AA. A national population-based study of incidence of colorectal cancer and age. Implications for screening in older Americans. *Cancer* 1995;75:775–81
13. Neugut AI, Forde KA. Screening colonoscopy. Has the time come? *Am J Gastroenterol* 1988;83:295–7
14. Lieberman D. Colon cancer screening: beyond efficacy [Editorial]. *Gastroenterology* 1994;106:803–12
15. Mandel JS, Church TR, Ederer R, Bond HJ. Colorectal cancer mortality: effectiveness of biennial screening for fecal occult blood. *J Natl Cancer Inst* 1999;91:434–7
16. Letsou G, Ballantyne GH, Zdon MJ, Zucker KA, Modlin IM. Screening for colorectal neoplasms. *Dis Col Res* 1987;30:839–43
17. Winawer SJ, Flehinger BJ, Schottenfeld D, Miller DG. Screening for colorectal cancer with fecal occult blood testing and sigmoidoscopy. *J Natl Cancer Inst* 1993;85:1311–18
18. Vernon SW. Participation in colorectal cancer screening: a review. *J Natl Cancer Inst* 1997;89:1406–22
19. Richardson JL, Danley K, Mondrus GT, Deapen D, Mack T. Adherence to screening examinations for colorectal cancer after diagnosis in a first-degree relative. *Prev Med* 1995;24:166–70
20. Triezenberg DJ, Smith MA, Holmes TM. Cancer screening and detection in family practice: a MIRNET study. *J Fam Pract* 1995;40:27–33
21. Lieberman DA. Cost-effectiveness model for colon cancer screening. *Gastroenterology* 1995;109:1781–90
22. Byers T, Gorsky R. Estimates of costs and effects of screening for colorectal cancer in the United States. *Cancer* 1992;70S:1288–95
23. Gupta TP, Jaszewski R, Luk GD. Efficacy of screening flexible sigmoidoscopy for colorectal neoplasia in asymptomatic subjects. *Am J Med* 1989;86:547–50
24. Gilbertsen VA, Nelms JM. The prevention of invasive cancer of the rectum. *Cancer* 1978;41:1137–9
25. Hertz REL, Deddish MR, Day E. Value of periodic examinations in detecting cancer of the rectum and colon. *Postgrad Med* 1960;27:290–4
26. Selby JV, Friedman GD. Sigmoidoscopy in the periodic health examination of asymptomatic adults. *JAMA* 1989;261:595–601

27. Newcomb PA, Norfleet RG, Storer BE, Surawicz TS, Marcus PM. Screening sigmoidoscopy and colorectal cancer mortality. *J Natl Cancer Inst* 1992;84:1572–5

28. Selby JV, Friedman GD, Quesenberry CP, Weiss NS. A case–control study of screening sigmoidoscopy and mortality from colorectal cancer. *N Engl J Med* 1992;326:653–7

29. Müller AD, Sonnenberg A. Prevention of colorectal cancer by flexible endoscopy and polypectomy. A case–control study of 32,702 veterans. *Ann Intern Med* 1995;123:904–10

30. Sarles HE, Sanowski RA, Haynes WC, Bellapravalu S. The long and short of flexible sigmoidoscopy: does it matter? *Am J Gastroenterol* 1986;81:369–71

31. Ransohoff DF, Lang CA. Sigmoidoscopic screening in the 1990s. *JAMA* 1993;269:1278–81

32. O'Brien MJ, Winawer SJ, Zauber AG, et al. The National Polyp Study: patient and polyp characteristics associated with high-grade dysplasia in colorectal adenomas. *Gastroenterology* 1990;98:371–9

33. Eddy DM. Screening for colorectal cancer. *Ann Intern Med* 1990;113:373–84

34. Atkin WS, Cuzick J, Northover JMA, Whynes DK. Prevention of colorectal cancer by once-only sigmoidoscopy. *Lancet* 1993;341:736–40

35. Rex DK, Lehman GA, Ulbright TM, et al. Colonic neoplasia in asymptomatic persons with negative fecal occult blood tests: influence of age, gender, and family history. *Am J Gastroenterol* 1993;88:825–31

36. Wu CS, Tung SY, Chen PC, Kuo YC. The role of colonoscopy in screening persons with family history of colorectal cancer. *J Gastroenterol Hepatol* 1995;10:319–23

37. Rozen P, Fireman ZVI, Figer A, Legum C, Ron E, Lynch HT. Family history of colorectal cancer as a marker of potential malignancy within a screening program. *Cancer* 1987;60:248–54

38. Rocklin MS, Slomski CA, Watne AL. Postoperative surveillance of patients with carcinoma of the colon and rectum. *Am Surg* 1990;56:22–7

39. Engstrom PF. Colorectal cancer. In Holleb A, Finn DJ, Murphy GP, eds. *Clinical Oncology*. Atlanta, GA: American Cancer Society, 1991

40. Winawer SJ, Fletcher RH, Miller L, et al. Colorectal cancer screening: clinical guidelines and rationale. *Gastroenterology* 1997;112:594–642 [Published errata appear in *Gastroenterology* 1997;112:1060 and 1998;114:625]

41. Hamilton SR. Colorectal carcinoma in patients with Crohn's disease. *Gastroenterology* 1989;89:205–16

42. Rozen P, Fireman ZVI, Figer A, Ron E. Colorectal tumor screening in women with a past history of breast, uterine, or ovarian malignancies. *Cancer* 1986;57:1235–9

43. Schapira DV, Pamies RJ, Kumar NB, et al. Cancer screening, knowledge, recommendations, and practices of physicians. *Cancer* 1993;71:839–43

44. Faivre J, Bader JP, Bertario L, et al. Mass screening for colorectal cancer: statement of the European Group for Colorectal Cancer Screening. *Eur J Cancer Prev* 1995;4:437–9

45. Byers T, Levin B, Rothenberger D, Dodd GD, Smith RA. American Cancer Society guidelines for screening and surveillance for early detection of colorectal polyps and cancer. *Cancer J Clin* 1997;47:157

46. Bogoch A. Bleeding from the alimentary tract. In Haubrich WS, Schaffner F, Berk JE, eds. *Bockus Gastroenterology*, 5th edn. Philadelphia, PA: W.B. Saunders, 1995

47. Shinya H, Cwern M, Wolf G. Colonoscopic diagnosis and management of rectal bleeding. *Surg Clin N Am* 1982;62:897–903

48. Richter JM, Christensen MR, Kaplan LM, Nishioka NS. Effectiveness of current technology in the diagnosis and management of lower gastrointestinal hemorrhage. *Gastrointest Endosc* 1995;41:93–8

49. Bat L. Medicine in the elderly – colonoscopy in patients aged 80 years or older and its contribution to the evaluation of rectal bleeding. *Postgrad Med J* 1992;68:355–8

50. Goulston KJ, Cook I, Dent OF. Diagnosis – how important is rectal bleeding in the diagnosis of bowel cancer and polyps? *Lancet* 1986;II:261–4

51. Giullem JG, Forde KA, Treat MR, Neugut AI, Bodian CA. The impact of colonoscopy on the early detection of colonic neoplasms in patients with rectal bleeding. *Ann Surg* 1987;206:606–11

52. Jensen J, Kewenter J, Asztely M, Lycke G, Wojciechowski J. Double contrast barium enema and flexible rectosigmoidoscopy: a reliable diagnostic combination for detection of colorectal neoplasm. *Br J Surg* 1990;77:270–2

53. Kewenter J, Brevinge H, Engaras B, Haglind E. The yield of flexible sigmoidoscopy and double-contrast barium enema in the diagnosis of neoplasms in the large bowel in patients with a positive hemoccult test. *Endoscopy* 1995;27:159–63

54. Williams R, Parker Davis I. Diverticular disease of the colon. In Haubrich WS, Schaffner F, Berk JE, eds. *Bockus Gastroenterology*, 5th edn. Philadelphia, PA: W.B. Saunders, 1995

55. Sohn N. Diseases of the anus, rectum, and sigmoid colon: the proctologiexamination. In Schapiro M, Lehman GA, eds. *Flexible Sigmoidoscopy*. Baltimore: Williams & Wilkins, 1990

56. Saltzman AK, Carter JR, Fowler JM, et al. The utility of preoperative screening colonoscopy in gynecologic oncology. *Gynecol Oncol* 1995;56:181–6

57. Rex DK, Mark D, Clarke B, Lappas JC, Lehman GA. Flexible sigmoidoscopy plus air-contrast barium enema versus colonoscopy for evaluation of symptomatic patients without evidence of bleeding. *Gastrointest Endosc* 1995;42:132–8

CHAPTER 2

Instruments for sigmoidoscopy

Brigitte E. Miller

Similar to the development in other fields of medicine, technology is advancing at a fast pace and significantly improved instruments become available every 5 years or so. However, the latest technology is not always essential. In this chapter the basic instruments needed for evaluation of the sigmoid, their maintenance and the set-up of an examination room are discussed. Prior to making the final purchasing decision, it is important to obtain several estimates from vendors.

RIGID INSTRUMENTS

Rigid instruments are rarely used in the office now, as flexible instruments are much more comfortable for patients. However, proctoscopes are necessary for complete evaluation of the anal canal. They come in varying diameters and lengths. The rigid sigmoidoscopes have a fiberoptic light system that runs circumferentially around the observation channel. The bowel mucosa is observed by direct vision. Biopsy forceps are available in a variety of designs.

Rigid instruments are easy to care for. They are mostly made of stainless steel. Cleaning with soap and water is carried out before they are autoclaved. Standard disinfection is sufficient. The sealing rings at the viewing area of the rigid scopes tend to become brittle and have to be changed occasionally. The rubber hoses and the blow bulb for air insufflation also tend to become porous and leak air. Otherwise, these instruments are extremely sturdy.

THE FLEXIBLE SIGMOIDOSCOPE

Flexible sigmoidoscopy was originally developed in Japan in the 1950s. The initial instruments introduced into the USA were still quite difficult to use and allowed only a limited view. This improved with progress in fiberoptic technology. Video endoscopes were then introduced in the early 1980s.

The basic components of a flexible sigmoidoscope are the insertion tip and the control head (Figures 1.1 and 1.2). The insertion tube and most of the control head and junction are covered by waterproof material, allowing for complete immersion during cleansing. Usually, an inner metal mesh is covered by an outer vinyl coat. The shaft is torque stable. The most distal centimeters, which are the most flexible, are also more prone to trauma, and therefore should be treated very carefully. Closer to the proximal junction to the control head, there is an inlet valve for the accessory or suction channel. In some instances, there is also an ancillary inlet for forceful water irrigation. Another frequent area of breakage and leakage is the connection site between the insertion tube and the control head.

The insertion tip includes the image system – either a lens system or an electronic charged coupled device (CCD) chip. In most instruments the view is at the tip. There are also instruments with a side view available, but these are necessary for only special procedures. The tip is slightly shielded with a protective hood to reduce soiling with stool or mucosal irritation during intubation. It also features the observation lens or electronic chip in addition to one or two light guides and the air/water and biopsy outlet. The lens is quite friable and may become chipped if not handled with care. There are usually three channel ports: one for air insufflation, one for water insufflation and a third for biopsy instruments or suction. The insertion tip also contains the cables for tip movement. Most instruments are capable of distal tip deflection of 180° in the up-and-down position and 160° in the right-and-left deflection. Four-way tip deflection is not always essential and can be achieved by up-and-down and turning maneuvers; however, it improves the ease of the examination and therefore is highly desirable. A full 180° deflection is especially important for complete evaluation of the rectum. A breaking system is incorporated, so that the tip can be kept in a special position.

The light is conducted to the tip through a fiberoptic bundle. The beam travels in a zigzag path reflecting off the surface of the fiber until it reaches the tip. The fibers are about 10 μm in diameter. Light-carrying bundles are not coherent. The core of the bundle has a very high refractive index for optimal light transmission. However, this leads to a greater absorption of the blue-spectrum light and therefore distortion of the colors. A compromise has to be made between the brightness and the spectrum. Observation is either through a coherent fiberoptic bundle or though an electronic system.

The water insufflation channel is manipulated by the air button on the control head. This is also coupled with the water irrigation function to clean the lens. Some cheaper models have separate ports for manual irrigation. This version is not as desirable but cheaper, and should be chosen only if cost is a major concern. Sometimes a syringe attachment has advantages, because the more forceful irrigation is more effective in cleaning the lens. The disadvantage is that the small attached syringe needs to be refilled several times during a procedure. On the other hand, a larger syringe is cumbersome. In addition, there is the risk of spillage.

The air channel is quite small. Air is supplied through the umbilical cord. There is no mechanism to check and control air pressure, therefore the examiner has to be careful not to overinflate the bowel which could lead to cecal injury. Some instruments deliver several different insufflation rates.

The channel for biopsy forceps has a minimum diameter of 2.6 mm; however, channels with a larger diameter up to 4.3 mm are also available. A larger channel allows not only better passage of a larger biopsy forceps but also improved aspiration of retained residue and improved flushing of the lens. The channel inlet is located close to the connection between the control head and the insertion tube but away from the observation lens, to reduce contamination by splashing. A rubber valve further reduces this risk. This has to be replaced not infrequently as it tends to deteriorate faster and then leakage can be a problem.

The main suction channel exits the control head to the umbilical cord, which then is attached to an external suction appliance or, even better, to a centralized suction system.

The second part is the control head, which includes the viewing and magnifying eyepiece and focusing ring on fiberoptic instruments or the exit port to the video processor. This area also contains the two control knobs for movement of the tip, as well as the buttons for water and air insufflation. The umbilical cord connects the control head to water, suction, light sources and video equipment.

Usually the proximal valve controls suction and the distal valve controls air insufflation when touched and in some instruments also water irrigation when depressed.

TYPES OF INSTRUMENT

A variety of different instruments are available. There are significant differences in price. Although some of the less expensive systems may be more prone to water leakage or be slightly more inconvenient to use, they may still be a valid alternative for offices where only a small number of examinations are carried out or as a back-up system. There is also a choice between fiberoptic and electronic video technology. Certainly the fiberoptic system is much cheaper; however, it is more inconvenient to use. With new technology, the image systems have become more trouble-free and reasonable in cost, even for the primary care physician.

Length

The average depth of insertion of a rigid sigmoidoscope is about 20 cm. Using a 35-cm flexible scope, a distance of about 30 cm is evaluated; with a 60-cm instrument, the distance increases to 50 cm[1]. The mean duration of the examination is slightly longer using the flexible scope, especially the 60-cm version. Although several studies have reached slightly different conclusions, most patients prefer the examination with the flexible scope[2,3]. In some reports, up to 50% of patients did not express a preference; however, most of those evaluations were performed in the 1970s. With better instruments and longer experience, the shift towards the flexible instruments is almost complete.

Flexible sigmoidoscopes have been compared with rigid sigmoidoscopes in a large number of studies, mainly in the 1970s and 1980s. Up to 65% of all colon polyps are within the reach of a 60-cm flexible sigmoidoscope. Compared with rigid sigmoidoscopy, an examination with the 60-cm flexible instrument increases the detection rate for polyps 3.1-fold and for cancer 3.6-fold[4]. The difference is even higher regarding diverticuli, which are rarely seen in the rectum or distal sigmoid. Only 2% of the lesions seen with the flexible scope were also seen using the rigid scope in one study. The differences regarding inflammatory bowel disease are less staggering, but still high: here 72% of the lesions seen using the flexible scope were also seen with the rigid scope[4]. The examination with the rigid scope can be very painful and difficult in the presence of significant inflammation and scarring.

Flexible sigmoidoscopes are available in lengths of 30–35 cm, 60–65 cm or 70–77 cm. The cost is not an issue; however, the longer flexible instruments may be slightly more complicated to use and the learning curve may be longer for the general practitioner. This does not mean that the motivated generalist cannot become proficient. Several studies attest to this. Once one is familiar with the 35-cm instrument, it takes only a few more sessions to become familiar with the 60-cm instrument. Again, the short flexible scope is superior to the rigid scope in all parameters, without significantly increasing the examination time. When the 35-cm instrument and the 60-cm instrument were compared, the examination time with the shorter instrument was shorter and patient tolerance was better. Overall, about 20 cm more of bowel lumen was evaluated with the longer instrument. The percentage of polyps diagnosed increased from 78% to 98%[5]. These results were supported by other reports. The diagnostic accuracy for inflammatory bowel disease was not different between the instruments. As expected, diverticular disease was diagnosed less frequently using the shorter instrument. However, the longer instruments result in a much better examination in that the entire sigmoid colon and even part of the descending colon can easily be examined. In almost 40% of cases the examiner overestimated the depth of insertion. Assessing flexible sigmoidoscopy by radiological methods, at full insertion of the 60-cm instrument the junction between the sigmoid and descending colon was reached in 70–94% of cases[4]. Full insertion was possible in 55–100% of the patients. These results depended significantly on examiner training and patient population, however. There are no comparable studies available about the complication rates of the short and the long instruments. A longer instrument may lead to a higher complication rate. As the overall complication rate is quite low, a prohibitive number of patients will have to be examined to achieve statistically significant results. There are few reports about instruments of 120 or 160 cm in length. In one publication only 3% of patients had a polyp detected above 60 cm without another polyp further downstream. Therefore, it seems that instruments longer than 60 cm lead to minimally increased diagnostic accuracy[6]. In addition, the longer the instrument, the more difficult the technique. Still, one has to keep in mind that even the 60-cm examination is no substitution for colonoscopy, as about one-third of the cancers and polyps are outside its reach. On the other hand, a longer segment of sigmoid examined increases the diagnosis of polyps in this area and thus the number of colonoscopies for complete evaluation of these patients.

In conclusion, the flexible 60-cm sigmoidoscope is the instrument of choice, owing to better diagnostic accuracy regarding polyps, cancer, diverticuli and inflammation when compared with rigid sigmoidoscopes and 30-cm flexible instruments, and to improved tolerance when compared with the rigid sigmoidoscope. Regarding the length of the flexible instrument, the longer instruments give a more accurate diagnosis, the price to pay being the slightly longer learning curve and slightly longer examination time. However, it is not worthwhile to go beyond 60 cm.

Diameter

The outer diameter of the sigmoidoscope can be either 12 or 16 mm. When smaller instruments are used, they are usually softer and more flexible, rendering the examination of the sigmoid colon more difficult, because of loop formation. The average depth of insertion is significantly less with a smaller-caliber instrument[7]. A more difficult examination takes longer and is more uncomfortable, therefore a smaller instrument does not translate into decreased patient discomfort. Smaller instruments should be reserved for patients with significant stenosis and should be used by specialists. For the general practitioner the larger size is recommended. In addition, a larger biopsy channel is desirable for improved tissue diagnosis.

Fiberoptics

The development of fiberoptic technology was a breakthrough for many medical instruments. These use bundles of 20 000 to 40 000 glass fibers to transmit light. The fibers are about 10 μm in diameter. In smaller fibers, too much light is lost during transmission, owing to diffraction. Each fiber is coated with a glass of a lower optical density to ensure total internal reflection inside the fiber and thus accurate light transmission. The light travels in a zigzag pattern. The pictures have a grid-like structure, owing to dead space between the fibers. The finer the bundles, the better the picture. These bundles are very flexible. For visualization the bundle has to be coherent, which means that the spatial relation of the fibers is stable so that an accurate picture is transmitted. In fiberoptic systems the differences in optical resolution and image reproduction are mainly related to the prominence of the fiber bundle patterns, angle of view and depth of field. Wide-angle lenses increase the field of vision and allow for easier identification of the bowel lumen in areas of flexures. On the other hand, they also lead to some distortion, such as overestimation of polyp size. This is rarely of clinical significance, as the observer easily becomes familiar with it. The depth of focus is fixed. The larger the depth of focus, achieved with a larger aperture, the less light gets transmitted. Therefore, the focus is fixed at a compromise between depth of focus and light intensity. The distal lens of the instrument is fixed and usually has a focal depth of about 3–15 mm. On the other end of the scope, the image can be visualized through a focusing lens with a magnifying effect up to 10×. Teaching attachments are available for fiberoptic scopes. These are often awkward to use. In addition, a brighter light source is necessary, as the light beam is split to supply both observation lenses. Only one additional person can watch. Video converters are available, which can be placed over the eyepiece and then transmit pictures to a monitor. However, the picture quality is slightly reduced.

The video endoscope is of the same basic design. However, the picture transmission is similar to television. A CCD chip is mounted at the tip, transmitting the picture through as many as 100 000 individual picture elements (pixels). If light hits the pixels, electrical signals are sent to the video processor and converted into an image. The color transmission follows the same principle as on a television. The color CCDs have extra pixels for an overlay of multiple primary color filters. A pixel under a particular stripe only responds to light of that color which is then transmitted. There are several technical ways to achieve this goal, all with good results. All signals are then further evaluated by computer and transmitted to a monitor. Overall, electronically transmitted pictures are clearer and of higher resolution. Initially colors were slightly artificial. However, with newer technology and better monitors, this disadvantage is no longer significant. In addition, the diagnostic accuracy depends more on identification of shapes and color differences rather than on exact individual colors. Good systems also produce adequate colors and have ample options for editing. Picture documentation is very easy with a video system. Image quality is high. The area from where the picture is taken can be easily documented. There is also an option to film the entire procedure if necessary for teaching purposes or quality control.

What is the better instrument, fiberoptic or video? The fiber endoscope is certainly much more awkward to handle, owing to the small lens used for viewing the pictures. In addition, the physician's face is closer to the instrument, increasing the risk from splashing. On the other hand, the colors are very accurate, but this may not be of clinical importance. Videoscopes can probably be made smaller than fiberoptic instruments and, with an increased number of pixels, the resolution is improved. However, with newer technology and better monitors, the colors are quite accurate. The video system renders the examination much easier, as the examiner can look at a video screen instead of the small eyepiece. Manipulation of the shaft can be much easier, if the eye does not have to follow every movement of the instrument head. In addition, the patient can also view the examination if so desired. Documentation of findings can be achieved with the push of a button, with no special preparation necessary. Teaching is very easy with a video system, as several people can watch the monitor. Fiberoptic instruments are simpler, but in most endoscopy suites, video equipment is used.

ACCESSORIES
Light sources

Again there is a wide variation of different light sources available. Low-cost halogen units are satisfactory for an examination; however, for photography they are not adequate. For good photographic resolution, a high-output xenon light unit is recommended. Again the best documentation can be accomplished using the electronic video systems.

Biopsy forceps

There are small brushes for cytology. Suction traps can be used to collect additional cytology samples. However, in most instances, a biopsy should be taken to obtain a histological sample.

A variety of different biopsy forceps are available. The diameter is obviously limited by the size of the instrument channel. A cup-biopsy forceps is the most versatile. Some forceps contain small needles to stabilize the tissue sample. A forceps with a serrated edge often cuts the tissue better. The sharper the forceps, the easier and more problem-free is the biopsy. Fenestrated biopsy cups often allow larger samples to be taken. Very large biopsy forceps are not recommended, because they require a larger biopsy channel; in addition, a larger biopsy also leads to a higher risk of complications, and cauterization may become necessary. A simple biopsy of a small lesion can be obtained safely without coagulation requirements. So-called hot biopsy forceps with cautery are not recommended for routine sigmoidoscopy. They are safe only if the colon has been completely cleaned. If a large polyp is seen, a complete colonoscopy is necessary, and the polyp is then removed after adequate colon preparation. Careful maintenance is of paramount importance. If biopsies are taken only rarely, single-use instruments should be considered[8].

Camera

Photographic documentation is a desirable feature. Using fiberoptic systems, this can be done with an attached camera, provided the light source is adequate.

With a video processor, photographic documentation is even easier and is done with the push of a button. While the photograph is taken, the depth of insertion and localization can be documented. Although this feature is more important for colonoscopy, it is also helpful for sigmoidoscopy if polyp resection may become necessary in a separate procedure.

WHAT TO BUY

Sigmoidoscopes are produced by several companies and there are a variety of systems to choose from. Considering the price, one has to estimate the number of examinations that will be carried out. For several weeks one should keep track of all patients who would benefit from sigmoidoscopy, symptomatic patients and patients above the age of 50 for screening. One has then to decide how many examinations can be done each day. Although the examination itself takes only 5–10 min, one has to allow at least 30 min to include set-up time, cleaning time, etc. Charges and reimbursements should also be evaluated.

If one expects to perform sigmoidoscopy on a regular basis, then it is reasonable to invest more in the instrument and look for convenience, such as a video system, for example. If, on the other hand, one intends to use the instrument only occasionally, such as for staging for cervical cancer, or as a back-up unit, then it is prudent to buy a simpler instrument. One should also try out different imaging systems to determine the most favorable. The small differences are not important for clinical use, but may make a difference regarding personal preference. Prices are overall comparable. However, it is worthwhile to order the entire set-up from one company, as adapters are then interchangeable.

In addition, it is very important to ensure that the company is responsive to all service requirements. Support services are a critical factor. Although the instruments are quite robust, a breakdown can always occur. Fast service, fast repair and replacement instruments are important, especially if scopes are needed on a regular basis.

STORAGE AND MAINTENANCE

Endoscopes are shipped in attractive suitcases. These containers should be kept in case shipment is needed later. These cases should not be used for storage. The foam padding increases the risk of infection and the curled position of the shaft increases the risk of breakage of the optic fibers. For the same reason, the instrument should not be stored in drawers. The best storage is hanging in a cabinet with the shaft in the straight position. Endoscopes are complex tools that need to be handled with care. They can easily be damaged if they are left to dangle. The quality of the maintenance significantly determines the durability of the instrument. The manufacturer's manual should be consulted.

Cleaning

Sigmoidoscopes are defined as semicritical medical devices, as they do not disrupt the mucosal membrane. High-level disinfection is adequate, killing all bacteria, viruses or fungi, but not necessarily all spores. Sterilization is not necessary. The Environmental Protection Agency publishes a list of recommended disinfectants as approved by the Food and Drug Administration, and can be viewed on the Internet. Biopsy devices transgress the mucosal membrane and therefore have to be sterilized, as they are regarded as critical devices. The risk of infection transmission after sigmoidoscopy is minimal. Only a small number of cases have been reported, although microbiological evaluation of instruments after proper cleaning and sterilization revealed bacterial contamination, mainly with *Pseudomonas* species in up to 24%[9]. Overall, the colony count numbers were small. Most viruses are removed with simple cleaning; the remainder are destroyed by glutaraldehyde. This holds true for hepatitis as well as HIV. *Clostridium difficile* is also mostly destroyed with routine cleaning procedures. Very few cases of possible transmission have been reported in the literature. Although *Mycobacterium* species are difficult to kill using the recommended soaking times, transmission has rarely been reported. If glutaraldehyde were used on its own, soaking times of up to 2 h would be necessary. However, in combination with vigorous mechanical cleaning, this seems not to be necessary. Prior to examination of severely immunocompromised patients, one may have to intensify the preparation. Transmission of *Salmonella* species during colonoscopy has been reported, but this is mainly due to incomplete processing. Most reported cases of contamination are related to the water sources used for rinsing and for automatic cleaning machines. Transmission from patient to patient is very rare.

Sigmoidoscopy is the endoscopy with the smallest risk for infection. Nevertheless, universal precautions should be routine. A clear protocol has to be written with detailed instructions regarding presoaking, manual cleaning, disinfecting, drying and storage. Especially in an office where these procedures are performed only occasionally, rigid guidelines are important, as there is a tendency towards incomplete cleaning. A summary document is published by the American Society of Gastroenterology as well as the Association for Professionals in Infection Control and Epidemiology (APIC) and is periodically updated. Cleaning efforts should be checked by cultures periodically. Only well-trained personnel should perform these tasks. It is not only the endoscope that needs to be properly disinfected; it is also important to keep the water bottle for irrigation free of bacterial contamination. If washing machines are used, these need to be regularly inspected to avoid contamination. The water supply should be monitored. In-line bacterial filters are strongly recommended. Cleaning solutions should be regularly checked for proper concentration.

Cleaning is important for safe use and durability of the instrument[10]. Good mechanical cleaning is a very important first step. This cannot be done automatically. The main reason for nosocomial infections is insufficient mechanical cleaning. All newer instruments can be completely submersed. Glutaraldehyde hardens organic matter, which then impairs complete exposure of all surfaces to the disinfecting agent and increases the risk that bacteria survive the disinfection. Immediately after use, the scope should be immersed in water and all channels should be flushed so that organic debris cannot dry, rendering it so much more difficult to remove. Enzyme detergents are especially useful for removing all organic debris outside and inside the channels. They should be used at room temperature for 5–10 min. A detergent, which lowers the surface tension as much as possible, is recommended. Household detergents are no substitute, as some contain lipids, which lead to a thin film covering all surfaces and render complete cleaning more difficult. It is important to ensure adequate concentration and dissolution of the detergent, especially if a powdered material is used. Small granules can clog small instrument channels. The following equipment is needed: large plastic basins or sink, warm water and soap solution, gauze pads, washcloths and gloves, cotton-tip applicators, 70% isopropyl alcohol, cleaning brush, towels, 20-ml syringe, glutaraldehyde or other disinfection solution.

Before cleaning, the instrument is carefully inspected for leaks, especially the soft bending section. This is best done by inflating the air channel under pressure. The shaft is immersed in warm water and washed with a soft cloth. The tip is cleaned with a soft brush. All channels are cleaned of debris by first being flushed with water and then having 100–200 ml of soapy water suctioned through all ports. All air should be expelled from the ports. Blockage of the air, water, or suction channel can occur frequently, mainly because of organic debris. First, it is important to check all connections, as well as the tightness of the rubber rings, wherever needed. Most often, one can clear the different channels by flushing them with a syringe. Water can be injected down any channel. A small syringe, 5–10 ml, generates more pressure than a large syringe. On the other hand, a larger syringe generates more suction. It is important to cover or depress the control button while syringing. Sometimes one has to remove the valve and apply suction directly to the port. Irreversible air channel blockage is mainly due to agglutinated residue inside. Sometimes, cleaning it with a small wire is necessary. The suction channel is cleaned with a long brush, which is passed through it entirely at least three times in both directions, towards the tip and towards the umbilical cord. A cotton-tip applicator is used to clean the channel inlets. The biopsy channel is cleaned the same way. The instrument is then rinsed in warm water. The channels are again washed out and air is pushed through to remove most of the water. All-channel irrigators are supplied by some companies to facilitate this aspect. If they are used, it is important to ensure that detergent appears from all ports. All removable valves are cleaned. Prior to being disinfected, the instrument should be dry, so that the disinfectant solution is not diluted. OSHA guidelines for the disinfecting room and ventilation should be followed.

Disinfecting should take place in a well-ventilated area, preferably under a hood. The solutions can cause significant mucosal irritation, respiratory problems and allergies. Thick rubber gloves should be worn, as well as a face mask or goggles to guard against splashing. Several different solutions are available. It is important also to consult the manufacturer's recommendation. Glutaraldehyde is mostly used, as it penetrates easily, owing to low surface tension. Alkaline glutaraldehyde is less corrosive than the acid preparation. It destroys most bacteria and viruses, including HIV and hepatitis, by alkylation of active groups, leading to destruction of nucleic acids and interruption of protein synthesis. A 2% solution should be used at a pH of 7.5 to 8.5[9]. As it tends to polymerize, one has to watch the shelf life, so that an active solution is used at all times. The disinfecting solution is suctioned through all channels. Although most of the effect against bacteria takes place within 4 min, the instrument should be soaked for 20 min at 20 °C to affect the viruses. Longer soaking is not much more effective, except if activity against spores is needed, but may impair the seals on the shaft. It is recommended, however, if there is possible exposure to mycobacteria or if the instrument is used in a severely immunocompromised patient. Overall, the risk of transmission of mycobacteria is minimal, even after routine soaking times. If the disinfecting is done manually, it is important to keep track of the numbers of use. A 2% glutaraldehyde solution is diluted to an ineffective 1% after 40 scopes have been cleaned. A 70% solution of isopropyl alcohol should not be used as a primary disinfectant, but is helpful in drying the instrument after disinfecting when used as a rinse after soaking. Washing machines are helpful and are used in most specialists' offices[11]. They may even be cost effective in a busy primary care office, where many sigmoidoscopic examinations are performed. In addition, the cleaning process is standardized. In no respect, however, do machines replace good careful manual cleaning. Machine lines have to be checked regularly for bacterial contamination. Also, one has to ensure that all channels are treated, especially the smaller air and water channels and not only the larger suction and biopsy channels. Sometimes special adapters are necessary to achieve this goal.

After disinfecting is completed, the instrument and channels have to be rinsed again in water and isopropyl alcohol. This also helps the drying process. Again, the channels have to be suctioned. The openings are cleaned with a gauze pad. It is very important to dry the channels completely, otherwise there is a higher risk of infection. The scope should then be connected to the air pump, and the air button completely depressed, so that any additional water is expelled. Finally, the head of the scope and the distal tip are cleaned with isopropyl alcohol.

Biopsy forceps are cleaned in the same way. Again, thorough mechanical cleaning is of paramount importance. This is the procedure that is most often carried out incompletely. The instruments should be dismantled as much as possible. If spiral forceps are used, these are often difficult to clean. An ultrasonic cleaner renders this much easier and more complete and is mandatory. As these instruments come into contact with the vasculature, they should be sterilized with ethylene oxide gas or in the autoclave. Increasingly, single-use items are available. This may be worthwhile, if biopsies are taken infrequently. Again, one should not coil the forceps too much, as this can damage the mechanics. The best way of storage is to hang them straight or bend to no more than one loop.

Cleaning of the water bottle is necessary, and it is also important to clean and sterilize water ports and tubes completely.

Cleaning and disinfecting of all instruments should be carefully documented. At least a tag with date and time of cleaning should be left. Additional documentation in a log book is also worthwhile. Cleaning should always be done as soon as possible after each examination, so that organic debris cannot dry, which makes it harder to remove. Instruments probably stay clean overnight. However, when examinations are not performed every day, then it is best to disinfect before the procedure, to ensure that the instrument is as clean as possible.

Maintenance

Good maintenance significantly prolongs the usefulness of the instrument; it starts with good cleaning and care. Before each examination all important functions should be checked. After each procedure the instrument should be checked for air leaks, by using an air pump or a special testing device. Thus, small leaks can be detected and repaired right away. Examination of the instrument under water is the next step. It is recommended that a maintenance person be identified to take care of these tasks and keep track of the instrument's performance. It is important to keep a log of all procedures done with a certain instrument, including dates and patients and examiners involved. Thus, a source can be traced in case of infectious complications, and poor-performing instruments can be identified. Close co-operation with the manufacturer and the local service person is also recommended. Sigmoidoscopes are among the most robust endoscopic instruments.

SET-UP

Endoscopy can be easily integrated into a general medical office. If more than five examinations are performed every day, a dedicated room is recommended[12]. A rectangular room, about 12 × 15 foot in size, is adequate (Figure 1.3). A toilet and changing room in close proximity are advantageous. Good ventilation is also necessary. If examinations are performed less frequently, a large examination room or treatment room is sufficient. Most of the equipment can be placed on a mobile unit. The examination table should be at the center. Any comfortable examination table can be used. Sometimes a tilt may be of advantage. A hydraulic table may render the examination more comfortable for the physician. The physician stands at the patient's right side. The instrument's umbilical cord exits at the physician's left side. A nurse stands on the physician's right side at the patient's feet or at the other side of the patient to help with instrumentation, especially when taking biopsies. The nurse should have access to a work area to prepare histological specimens if necessary. The work zones of the endoscopist and assistant should be separated from each other. Adequate counter space is important. A cabinet should be available in a central location to store all supplies in easy reach of the nurse. The video monitor is located on the other side of the patient, in easy view for the examiner. For more complicated procedures and when a two-person examination technique is used, a second monitor behind the examiner in easy view of the assistant is also worthwhile. Although the risk of complications is very small, adequate emergency equipment should be available.

It is important to keep the examination room as comfortable as possible. Most of the equipment needed should be stored in cabinets so that the patient is not frightened by all the technical instruments.

As for all procedure rooms, adequate lighting is important. It is also of advantage if the light can be dimmed during the examination. A communication system as well as an emergency light system are recommended. Strong ventilation is important, as mentioned above. Centralized suction is preferred, but in smaller office buildings a separate system may have to be used. It is important to avoid contamination while cleaning the system. Compressed air should be available wherever the scopes are cleaned. An additional water supply is necessary when a washing machine is installed. Although rarely needed for sigmoidoscopy, an oxygen supply is advantageous.

For cleaning and disinfecting, a separate area of the room with a large sink and good ventilation should be available. Manual cleaning immediately after the procedure is important, even when a washing machine will be used for final cleaning. A separate room is even better (Figure 1.4), especially when the instruments are cleaned by hand. A fume hood with extra strong ventilation will further decrease the risk of exposure to irritating fumes from detergents and disinfecting agents, which otherwise may cause significant skin irritation and allergies. A large work space is recommended, which can be separated into dirty and clean areas. The size also depends on the number of procedures done each day. If many examinations are done each day, washing machines are worthwhile. In addition, they can be placed in the examination room without much trouble.

CONCLUSIONS

(1) A 60-cm scope is recommended.

(2) A video system is the preferred version, as it renders the examination and documentation easier.

(3) Mechanical cleaning is the most important part of instrument maintenance and infection prevention, and should be carried out with the utmost care.

(4) Strict rules should be set up and enforced for disinfecting.

(5) Identification of a staff member responsible for maintenance of all involved instruments is recommended.

(6) A dedicated room is more convenient; however, all instruments can be placed on a mobile cart and a large examination or treatment room is also adequate.

REFERENCES

1. Marks G, Boggs W, Castro AF, et al. Sigmoidoscopic examination with rigid and flexible fiberoptic sigmoidoscopes in the surgeon's office: a comparative prospective study of effectiveness in 1,012 cases. Dis Colon Rectum 1979;11: 162–8

2. O'Connor JJ. Flexible sigmoidoscopy: is it of value? Am Surg 1979;45:647–8

3. Manier JW. Fiberoptic pansigmoidoscopy: an evaluation of its use in an office practice. Gastrointest Endosc 1978;24:119–20

4. Katon RM, Keeffe EB, Melnyk CS. Flexible Sigmoidoscopy. Orlando: Grune & Stratton, 1985:77–81

5. Durbow RA, Katon RM, Benner KG, et al. Short (35 cm) versus long (60 cm) flexible sigmoidoscopy: a comparison of findings and tolerance in asymptomatic patients screened for colorectal neoplasia. Gastrointest Endosc 1985;31:305–8

6. Schuman BM, McKay MD, Griffin JW Jr. The use of the 130-cm colonoscope for screening flexible sigmoidoscopy. Gastrointest Endosc 1988;34:459–60

7. Hawes RH, Lehman GA, O'Connor KW, Kopecky KK, Lappas JC. Effect of instrument diameter on the depth of penetration of fiberoptic sigmoidoscopes. Gastrointest Endosc 1988;34:28–31

8. Sandler RS, Cummings MS, Keku TO, Terse A, Mehta N. Disposable versus reusable biopsy forceps for colorectal epithelial cell proliferation in humans. Cancer Epidemiol Biomarkers Prev 2000;9:1123–5

9. Alvarado CJ, Reichelderfer M. APIC guidelines for infection prevention and control in flexible endoscopy. Am J Infect Control 2000;28:138–55

10. Ott BJ, Gostout CH. Endoscope maintenance and repairs. Gastrointest Endosc Clin N Am 1993;3:559–69

11. Bradley CR, Babb JR. Endoscope decontamination: automated vs. manual. J Hosp Inf 1995;(Suppl):537–42

12. Gostout CJ, Ott BJ, Burton D, et al. Design of the endoscopy procedure room. Gastrointest Endosc Clin N Am 1993;3: 509–25

CHAPTER 3

Patient preparation for sigmoidoscopy

Brigitte E. Miller

Sigmoidoscopy can easily be included in a routine office visit, especially as a screening examination. Nevertheless, appropriate preparation is necessary. Thorough patient education allows for an easier examination and lowers the risk of problems or complications. This is time well spent.

MEDICAL HISTORY

A complete and detailed history is appropriate. It is particularly important to document previous gastrointestinal complaints, procedures and diagnoses. Detailed questions should be asked about bowel function, any change in bowel habits, as well as the presence of rectal bleeding, constipation, chronic diarrhea, painful bowel movements, or tenesmus. One should also inquire about abdominal symptoms such as distension, bloating, pain, cramping, decreased appetite, early satiety and nausea or vomiting, as well as general symptoms such as decreased energy level or fever.

Pre-examination evaluation also should include a short general medical history to detect significant problems, such as a history of myocardial infarction, endocarditis, mitral valve prolapse, the presence of a prosthetic heart valve, a new vascular graft, an orthopedic prosthesis, coagulation disorders, or diabetes. One should inquire about previous pelvic or abdominal surgical procedures. It is also important to be aware of immunosuppression or HIV infection.

Medication history is relevant, especially regular use of aspirin, non-steroidal anti-inflammatory drugs (NSAIDs), anticoagulants (all may cause increased risk of bleeding after biopsy), iron (renders bowel preparation more difficult), steroids, immunosuppressive drugs (may blunt an inflammatory reaction), antidepressants, or monoamine oxidase (MAO) inhibitors (if conscious sedation is used). One should also check whether the patient is using laxatives on a chronic basis (look for melanosis coli).

SPECIAL CONSIDERATIONS

The very anxious patient

Sigmoidoscopy is frequently a short procedure, so sedation is often unnecessary. Careful explanation of the procedure and the sensations the patient will experience will reduce anxiety. Successful, uncomplicated procedures take place in a calm, controlled environment with caring and supportive staff.

Rarely, a patient may be able to co-operate only with premedication. Oral medication with diazepam or oxycodone hydrochloride was not effective in a placebo-controlled trial[1]. The onset of the effect is too slow and the duration too long for a procedure, which takes about 10 min. Short-acting intravenous medication is more effective, but requires close monitoring during and after the procedure, as well as continuous pulse oximetry. A full resuscitation kit, equipped with antagonist medications such as naloxone (Narcan®) and flumazenil (Romazicon®) should be immediately available. Iber and colleagues[2] noted an overall complication rate of 7% in over 500 patients undergoing a variety of endoscopies under conscious sedation. The most common complication was decreased oxygenation. Frail or elderly patients (age > 60 years), those with multiple medical problems (cardiac/pulmonary) and those undergoing a prolonged examination will experience complications more frequently. The risk of serious complications necessitating further therapy was only 0.5% in a survey among general gastroenterologists[3]. The preferred sedative agent, which also gives significant amnesia, is midazolam (Versed®) at a dose of 0.03 mg/kg intravenously given over 2 min[4]. In elderly or frail patients, especially those with chronic liver disease, a total dose of 1.5 mg or less may be adequate. In combination with opiates, 1 mg of midazolam is often sufficient. There are other options: diazepam (Valium®) is also useful at a dose of 0.1 mg/kg (problems here involve vein irritation and less amnesia), meperidine

(Demerol®) at a dose of 25–50 mg[5], low-dose fentanyl (Alfenta®) 50–100 μg intravenously immediately prior to the procedure. Close monitoring during and after the procedure is imperative. MAO inhibitors should be discontinued for at least 14 days.

The elderly patient

The percentage of older patients continues to increase. There are more abnormal findings in this patient population and the risk of colon cancer increases significantly with age. Therefore, evaluation of symptoms and general screening is very important in this group. Increased complication rates due to the effects of bowel preparation such as dehydration and a higher risk of hypoxia if conscious sedation is used during the procedure have been reported, but could not be confirmed in recent comparative studies[6]. Bowel preparation was more often unsatisfactory in older patients, owing to chronic constipation and reduced compliance. The completion rate for colonoscopy (complete visualization all the way to the cecum) is also lower, often because of diverticulitis. In a report of over 300 colonoscopies, however, there were no severe medical complications. Complication rates for sigmoidoscopy are even lower. In conclusion, colonoscopy and especially sigmoidoscopy can be safely performed in the elderly[7].

The pregnant patient

Rarely, flexible sigmoidoscopy may be necessary during pregnancy to evaluate severe diarrhea or rectal bleeding. In a small report on 24 patients, no complications were noted[8]. Likewise, there was no evidence that labor was induced as a result of sigmoidoscopy.

Prevention of bacterial endocarditis

Introducing instruments into the gastrointestinal tract has not proved to be a frequent means of entry for organisms that cause bacterial endocarditis. Although 10% of patients have been noted to have bacteremia after rigid sigmoidoscopy, transient bacteremia without clinical sequelae was noted in only one of 100 patients in a more recent study using a flexible sigmoidoscope[9]. Bacteremia is mainly by Gram-negative organisms and is rarely responsible for endocarditis.

Guidelines from the American Society for Gastrointestinal Endoscopy in conjunction with the American Heart Association (AHA) are based on minimal available data. Antibiotic prophylaxis is recommended for endoscopic procedures performed on patients with prosthetic heart valves, a history of endocarditis, or a surgically constructed systemic pulmonary shunt[10]. Mitral valve prolapse is the most commonly reported cardiac lesion, accounting for 49% of all cardiac risk factors. Traditionally, antibiotic prophylaxis prior to sigmoidoscopy was suggested for these patients if valvular regurgitation was present[11]. The newest AHA guidelines now state that prophylactic antibiotics are optional for patients with mitral regurgitations, other types of valvular dysfunction, or significant cardiomyopathy. There is no evidence that coronary artery bypass graft surgery increases the risk of endocarditis. Similarly, there are no definite recommendations for patients after heart transplants.

The most common non-cardiac risk factors are a new prosthetic joint or a vascular graft, but infection is very rare in these patients. Except for an occasional case report, no increased risks have been noted[12]. In patients with severe neutropenia or significant immunosuppression, the decision regarding antibiotics has to be made on an individual basis.

Antibiotics that should be used in high-risk patients are: ampicillin 2 g intravenously or intramuscularly and gentamycin 1.5 mg/kg intravenously 30 min prior to the procedure, with a second 1.5-g dose of ampicillin 6 h later. In penicillin-allergic patients, vancomycin 1 g can be substituted. Alternatively, an oral regimen including amoxicillin 3 g prior to the procedure and 1.5 g 6 h later can be used.

After myocardial infarction

Performing sigmoidoscopy after a myocardial infarction is safe in clinically stable patients[13]. However, the risk of cardiac complications was increased in unstable patients. Therefore, a recent myocardial infarction is not an absolute contraindication for sigmoidoscopy. If possible, though, the examination should be delayed for 6 weeks after an acute event. If sigmoidoscopy is used for screening only, then 6 months should be allowed to pass after the acute event. Endoscopy soon after myocardial infarction should be performed with electrocardiography and pulse oximetry monitoring; in addition, one should avoid excessive air insufflation.

The anticoagulated patient

During sigmoidoscopy, the anticoagulated patient is at increased risk of bleeding from mucosal abrasions or, rarely, from bleeding into the mesentery. If a biopsy is taken, bleeding can occur acutely or be delayed up to 14 days later. For patients on anticoagulant therapy, the risk of bleeding has to be compared with the risk of recurrent thromboembolism. Overall, the risk of bleeding is low during uncomplicated screening sigmoidoscopy; of course, it is higher for a difficult procedure with biopsy or in the presence of significant inflammation. It is difficult to interpret the literature because most reports deal with the perioperative management of patients on chronic anticoagulation[14], and therefore are not fully applicable to sigmoidoscopy. For most procedures, it is adequate to discontinue warfarin for 2–3 days and proceed if the International Normalized Ratio is normal. Patients at very high risk for thromboembolism then need heparin therapy, either subcutaneously or intravenously. The medication is stopped 6 h prior to the procedure and restarted immediately thereafter. After an uncomplicated examination, warfarin may be restarted the next day. If biopsies were taken or polyps removed, one may wait several days before resuming oral anticoagulation.

Since chronic aspirin use also increases bleeding risk, it should be discontinued 1 week prior to the procedure. Reported complications were mainly related to biopsies[15]. Other drugs that can affect the coagulation system, such as NSAIDs, Persantin® or Trental®, can be continued before and after sigmoidoscopy.

The diabetic patient

Sigmoidoscopy can usually be performed without diet modification after cleansing enemas. Sometimes, however, more extensive preparation, including a clear liquid diet the evening before and nothing per mouth after midnight, may become necessary. In this case, the examination should be scheduled early in the morning so that food and medication can be resumed without unnecessary delay. In patients with brittle diabetes, modification of insulin dosage and frequent glucose monitoring may become necessary. Pre-procedure blood glucose monitoring by the finger stick colorimetric method is recommended in all diabetics. Hypoglycemia, even if asymptomatic, should be corrected before proceeding with flexible sigmoidoscopy. This will not affect the bowel preparation and will facilitate an uncomplicated examination.

Monitoring

Vital signs should be taken before and after the procedure in all patients. Continuous pulse oximetry is recommended for patients undergoing colonoscopy; however, as sigmoidoscopy is much shorter, it is rarely necessary except in frail patients and patients with significant cardiopulmonary problems. Whenever conscious sedation is used, close monitoring is imperative.

PHYSICAL EXAMINATION

Routine vital signs are taken prior to the procedure, along with a short interval history. An abdominal examination should then be carried out to rule out signs of inflammation, peritonitis, or perforation.

LABORATORY EVALUATION

In healthy patients, especially those undergoing a screening examination, no pre-examination laboratory evaluation is necessary. Patients with a significant amount of rectal bleeding should have the hemoglobin/hematocrit evaluated. If an acute abdominal problem is suspected or there is severe inflammation, a complete blood count, as well as other tests, should be obtained as indicated. Patients receiving chronic anticoagulation should have a current prothrombin time, partial thromboplastin time and platelet count.

BOWEL PREPARATION

Preparation for sigmoidoscopy is an uncomfortable but important component of a successful procedure. To maximize compliance with a screening or surveillance program, it is important to develop an easy, well-accepted bowel preparation.

Usually, local measures are adequate for preparation of the distal colon. The most frequently used method is a hypertonic phosphate enema (Fleet® enema) 1 h before the procedure. Several studies have shown that two enemas, or even more, do not improve outcome[16]. With this simple regimen, adequate preparation can be achieved in 80% of patients. Rare complications from using Fleet enemas have been reported, including hyperphosphatemia and hypocalcemia[17]. Usually the serum phosphor values rise slightly, but remain within normal limits and return to pretreatment levels within 10 h[18]. There are, however, several reports of life-threatening electrolyte imbalances. Patients with renal insufficiency, those reporting a long retention time, and children are at higher risk. Hyperphosphatemia has also been seen in patients with severe ulcerative colitis. Fleet enemas may also influence the colon mucosa, causing surface epithelial sloughing, leading to mild inflammatory changes on endoscopic and histological examination. Normal saline enemas (130 ml) can be used if the detection of minor inflammatory changes is the primary objective. They are as effective as the Fleet enema with no risk of serum electrolyte abnormalities or change in histological appearance. Tap water enemas are much less effective in cleaning the colon, although the histological appearance is not affected. Careful, stepwise instructions should be given to all patients, explaining their bowel preparation. The majority of patients are willing and able to complete this successfully at home. In special cases, administering the preparation in the office may become necessary.

Suppositories are not an important component of a sigmoidoscopy preparation. Used alone, suppositories are less effective and they are not necessary after an enema or bowel lavage[19].

Oral preparations with sodium phosphate, castor oil, or large-volume gut lavage using polyethylene glycol (PEG) electrolyte solutions is rarely necessary. These methods are used primarily for patients undergoing full colonoscopy and prior to sigmoidoscopy, when a barium enema is also scheduled[20]. In addition, some patients may prefer these oral preparations[21]. Since oral sodium phosphate may cause a reduction in intravascular volume, it should be used cautiously in patients with congestive heart failure, ascites, or renal insufficiency. In the presence of inflammatory bowel disease, absorption can be increased. Finally, sodium phosphate should not be used in pregnant or breast-feeding women, because the placenta actively transports phosphates and they are secreted into breast milk as well. Castor oil is used less frequently, because of the potential for abdominal cramping or premature labor in pregnant women.

For colonoscopy, bowel irrigation with PEG electrolyte solutions is the preferred preparation technique (using GoLytely®, Colyte®, Nulytely®). Although it is sometimes difficult for the patient to drink the large amount of fluid (one gallon), most tolerate this preparation without major electrolyte imbalances. Adding some ice may make it more palatable; however, the effectiveness is decreased when it is diluted too much. In some reports, administration of a prokinetic agent such as metoclopramide (Reglan®) was noted to facilitate gastric emptying and small bowel transit time and decrease nausea. In other reports, no influence was seen. There is no influence on the colon mucosa. Smaller volumes of electrolyte solutions (2 l) have been recommended for sigmoidoscopy. When this technique is used, enemas or suppositories are not necessary.

Special considerations

Medications

One should advise the patient to discontinue all oral iron medication for 2 weeks prior to the examination, as iron can make the feces adhere to the bowel wall, leading to mucosal coating and subsequent incomplete evaluation. Other medications that slow bowel function, such as codeine, should also be discontinued if possible. It is also recommended to discontinue a high-fiber diet for a few days prior to the examination.

Constipation

Patients with a history of severe constipation may have more problems getting adequately prepared for sigmoidoscopy. These patients should increase their laxative dose for several days so that a soft stool is formed without causing diarrhea. Local preparation should be completed with enemas.

Inflammatory bowel disease

There is some concern that aggressive bowel cleansing can aggravate inflammatory bowel disease. If the examination is performed in a patient with an established diagnosis of inflammatory bowel disease, a routine preparation can be used. Saline enemas are preferred, because they have less effect on the sigmoid mucosa. In patients with active disease or diarrhea, the examination may have to be performed without preparation, so that cultures can be obtained and subtle mucosal changes better seen.

The elderly patient

Elderly patients tend to have more discomfort with bowel preparation. As mentioned above, however, advanced age has not been shown significantly to influence the adequacy of preparation if additional problems such as constipation are taken into consideration. In general, enemas are better tolerated than lavage or cathartic methods.

The pregnant patient

Preparation for routine sigmoidoscopy with local measures does not influence pregnancy. However, aggressive cathartics should be avoided in this setting.

Special procedures

Regional colon preparation is adequate for the vast majority of examinations, even if a biopsy is taken. However, polypectomy should be performed only after complete and documented colon preparation, to prevent explosion of combustible gases if electrosurgery is required.

PATIENT INSTRUCTIONS

Well-informed patients are generally more co-operative. A thorough discussion of the indications, risks, complications and alternatives to the examination is essential. All patient questions should be answered completely and each step of the procedure should be clearly described, including possible pain or discomfort. Written information that the patient can take home and review may help reinforce this teaching. Immediately prior to the procedure, the physician should again ensure that there are no further questions. We use the form shown as Table 1.

CONSENT

The most important factor in obtaining the consent is a thorough discussion of the procedure, its indications,

PATIENT PREPARATION FOR SIGMOIDOSCOPY

Table 1 Patient instructions for sigmoidoscopy

During sigmoidoscopy a flexible tube will be placed into the lower part of your large bowel to examine it for abnormalities. To better visualize the bowel wall, gas will be insufflated. Sometimes a small tissue sample will be taken.

Preparation

Please tell your doctor about other medical problems (especially heart and lung conditions or diabetes), medications you are currently taking (especially blood thinners) and any drug allergies. Do not take aspirin or iron medications for 2 weeks prior to the procedure.

It is very important that your bowel is well cleaned. This makes a complete examination possible. Please follow the instructions marked.

(1) Use a Fleet enema to clean your bowel immediately before you come to the office. For best results, try to retain it as long as possible. The enema may be repeated in the office to complete the prep.

(2) The day before the examination, mix Colyte® with water as directed. Make sure that all granules are dissolved. Drink all between 2:00 p.m. and 6:00 p.m., consuming at least an 8-ounce glass every 15 minutes. Adding a small amount of ice will make it more palatable; however, too much ice renders it less effective. If you experience nausea, drink slower. About 1.5 h after this preparation is started, bowel movements will begin and the nausea often improves. Resume drinking the fluid until the bottle is empty. Your bowel movements should be clear at the end of the preparation. Only clear liquids are allowed after the prep. Do not eat or drink anything after midnight. Take your regular medications with a sip of water. If you take insulin or other diabetic medications, ask your doctor for specific instructions.

Examination

When you come to the office, your nurse or doctor will explain the procedure again. Please feel free to ask questions and inform them of any new complaints you may have. Your vital signs will be taken and then you will be placed as comfortably as possible on the examination table on your left side. You will be asked to bend your hips and knees. First the doctor will do a rectal examination and then the scope will be gently inserted. This may cause some mild discomfort. For better visualization, gas will be insufflated into your bowel. This may cause some cramping. It is temporary and will resolve once the gas is removed. This examination may cause some discomfort, but it should not hurt. Inform your doctor if you are in pain. A sigmoidoscopy will take about 15 min to complete.

After the examination

After the examination, you will be asked to expel all gas. You may then resume normal activities and diet. Most patients will not need special premedication to undergo this examination. If pain medication is necessary in your case, then you will be asked to rest for an hour after the procedure and resume your diet slowly. In this case, you should also not drive a car, work, or make important decisions during the remainder of the day.

If you had a biopsy taken during the procedure, it will be sent to the pathologist for further evaluation. We will notify you of the result as soon as possible. If you have not received a call within one week, please call us. You may notice mild rectal bleeding for a few bowel movements. Please avoid heavy lifting and straining for one week and avoid medications containing aspirin for one week.

Complications

Although sigmoidoscopy is a very safe procedure, complications such as severe bleeding, a tear or other injuries of the colon, severe blood pressure changes or adverse reactions to medications are possible. All these events are rare, occurring in about 1 in 1000 procedures. If a complication occurs, emergency surgery is sometimes necessary. Therefore, it is very important that you notify your physician immediately if you experience severe rectal bleeding or pain, black stools, nausea or vomiting, fever, fainting, or inability to have a bowel movement for more than two days.

If you have any more questions feel free to call ....

In case of emergency call..

Table 2 Consent form

Name: _____ Date: _____ Time: _____ am/pm

1. I authorize Dr. _____ to perform a <u>sigmoidoscopy</u> with biopsies of the bowel wall if indicated.
2. The purpose of the examination has been explained to me and is

   ☐ to screen for colon cancer,   ☐ to evaluate bowel problems.

3. The risks involved and possible complications have been explained to me, including but not limited to severe bleeding, injury to the bowel, emergency surgery and other medical problems. No guarantee or assurance has been given by anyone as to the results that may be obtained.
4. Alternative methods have been explained to me.
5. I give my consent to treatments and medical procedures on my body by all qualified personnel working under the supervision of the aforementioned doctor, before, during and after the procedure to be performed.
6. I consent to the disposal of any tissue which may be removed.

Signed: _____ Relationship: _____ Date: _____

I have given to the above individual an explanation of the contemplated procedure and its anticipated benefit, risks and potential complications.

Signed: _____ MD   Date: _____

I certify that I was present at the time the above explanation was given and, in my opinion, the subject understands the factors of this consent. I also witnessed the signatures of both parties.

Signed: _____ Position: _____ Date: _____

---

risks, complications and alternatives with ample time to answer the patient's questions. This is an excellent opportunity to develop a good relationship with the patient, to learn more about concerns or fears and to get to know the patient personally. The explanations should be tailored to the patient's vocabulary, education and personality. A patient-oriented disclosure requires discussion of all facts that are necessary to enable a lay person to make an informed decision regarding the examination. Even though the complication rate is very low, problems such as bleeding, bowel injury, or perforation requiring surgery should be mentioned. Patient participation should be encouraged and the patient should be regarded as a partner. The best time to inform the patient is during an office visit before the scheduled procedure, allowing sufficient time for the patient to consider the examination. A scared and nervous patient may not listen and completely understand your discussion immediately prior to the examination. It is recommended to have a witness present at the time of the explanation. Final documentation includes the signing of a consent form. A special form, which includes the information discussed, is better than a generic consent. An example is printed in Table 2.

CONCLUSIONS

Adequate patient preparation for sigmoidoscopy includes the following steps:

(1) Counseling regarding indications, risks and complications;
(2) Explanations about the course of the examination, including possible discomfort;
(3) Review of the past medical and medication history;
(4) Cleaning of the sigmoid colon with enemas;
(5) Abdominal physical examination;
(6) Consent form.

REFERENCES

1. Pound AC, Brown ED, Yawes RH, O'Connor KW, Lebinan GA. Oral sedative/analgesic premedication for 60 cm fiberoptic sigmoidoscopy. *Gastrointest Endosc* 1989;35:70–1
2. Iber FL, Sutberry M, Gupta R, Kruss D. Evaluation of complications during and after conscious sedation for endoscopy using pulse oximetry. *Gastrointest Endosc* 1993;39:620–5
3. Keeffe EB, O'Connor KW. 1989 A/S/G/E survey of endoscopic sedation and monitoring practices. *Gastrointest Endosc* 1990;36:S13–S18

4. Kallar SK. Conscious sedation in ambulatory surgery. *Anesth Rev* 1991;18(Suppl 1):9–12

5. Clvoklvavatia S, Nguyen C, Williams R, et al. Sedation and analgesia for gastrointestinal endoscopy. *Am J Gastroenterol* 1993;88:393–6

6. Ure T, Dehghan K, Vernave AM III, Longo WE, Andrus CA, Daniel GL. Colonoscopy in the elderly, low risk, high yield. *Surg Endosc* 1995;9:505–8

7. DePrima RE, Barkin JS, Blinder M, Goldberg RI, Phillips RS. Age as a risk factor in colonoscopy: fact versus fiction. *Am J Gastroenterol* 1988;83:123–5

8. Cappell MS, Sidhom O. Multicenter, multiyear study of safety and efficacy of flexible sigmoidoscopy during pregnancy in 24 females with follow-up of fetal outcome. *Dig Dis Sci* 1995;40:472–9

9. Goldman GD, Miller SA, Furman DS, Brock D, Ryan JL, McCallum RW. Does bacteremia occur during flexible sigmoidoscopy? *Am J Gastroenterol* 1985;80:621–3

10. Zuckerman GR, O'Brien J, Halsted R. Antibiotic prophylaxis in patients with infectious risk factors undergoing gastrointestinal endoscopic procedures. *Gastrointest Endosc* 1994;40:538–43

11. Dajani AS, Bisno AL, Chung KJ, et al. Prevention of bacterial endocarditis. Recommendations by the American Heart Association. *JAMA* 1990;264:2919–22

12. Norfleet R. Infective endocarditis and infections of orthopedic and vascular prostheses following gastrointestinal endoscopy and dilatation. *Gastrointest Endosc* 1990;36:546–7

13. Cappell MS. Safety and clinical efficacy of flexible sigmoidoscopy and colonoscopy for gastrointestinal bleeding after myocardial infarction. *Dig Dis Sci* 1994;39:473–80

14. Stein PD, Alpert JS, Copeland J, et al. Antithrombotic therapy in patients with mechanical and biological prosthetic heart valves. *Chest* 1992;102(Suppl):445S–455S

15. Dyer WS, Quigley EMM, Noel SM, et al. Major colonic hemorrhage following electrocoagulating (hot) biopsy of diminutive colon polyps: relationship to colonic location and low dose aspirin therapy. *Gastrointest Endosc* 1991;37:361–4

16. Preston KL, Peluso FE, Goldner F. Optimal bowel preparation for flexible sigmoidoscopy – are two enemas better than one? *Gastrointest Endosc* 1994;40:474–6

17. Biberstein M, Parker BA. Enema-induced hyperphosphatemia. *Am J Med* 1985;79:645–6

18. Cohan CF, Kadakia SC, Kadakia AS. Serum electrolyte, mineral, and blood pH changes after phosphate enema, water enema, and electrolyte lavage solution enema for flexible sigmoidoscopy. *Gastrointest Endosc* 1992;38:575–8

19. Lee MG. Comparison of three bowel preparations for sigmoidoscopy. *WI Med J* 1993;42:118–20

20. Kolts BE, Lyles WE, Achem SR, Burton L, Geller AJ, MacMath T. A comparison of the effectiveness and patient tolerance of oral sodium phosphate, castor oil, and standard electrolyte lavage for colonoscopy or sigmoidoscopy preparation. *Am J Gastroenterol* 1993;88:1218–23

21. Hickson DEG, Cox JGC, Taylor RG, Bennet JR. Enema or Picolax as preparation for flexible sigmoidoscopy? *Postgrad Med J* 1990;66:210–11

CHAPTER 4

The sigmoidoscopy examination

Brigitte E. Miller

Manual dexterity is an important factor in performing fast, painless and complete sigmoidoscopy. Only practice makes perfect. A good understanding of the basic anatomy and its variations is very helpful. In addition, one has to be familiar with the mechanics of the sigmoidoscope and the basic maneuvers necessary to negotiate the winding bowel lumen.

ANATOMY OF THE LOWER GASTROINTESTINAL TRACT

Colon

The anatomy of the colon relevant to flexible sigmoidoscopy and colonoscopy is extremely variable and difficult to predict from radiological studies. Both the shape of the pelvis and the length of the mesentery, as well as pericolonic adhesions and uterine or adnexal masses, cause variations. Using intraoperative measurement, Saunders and colleagues[1] found the mean colon length to be 114 cm. The length of the rectosigmoid was reported as 34 cm with a range between 17 and 78 cm. The average descending colon length was 18 cm with a range between 11 and 42 cm. The amount of air insufflation at the time of endoscopy has an important effect on the total colonic length. The sigmoid mesocolon, which determines the mobility of the sigmoid, can be highly variable, ranging between 2 and 26 cm, with an average length of 11 cm. In about 17% of patients, the sigmoid is not mobile, owing to adhesions. The descending mesocolon is much shorter, about 4.8 cm in median length. A long, mobile mesentery may render examination more difficult, owing to loop formation. A short mesentery may render the examination painful to the patient if care is not taken to avoid excessive tension during advancement of the endoscope. Blood supply for the sigmoid and upper rectum originates from the inferior mesenteric artery, with anastomoses to branches from both internal iliac arteries supplying the lower rectum and anal canal.

Anus

The anal canal is about 3–4 cm long and points towards the umbilicus. The lower portion is lined with squamous epithelium. The anal papillae mark the dentate or mucocutaneous line 2 cm from the anal verge. Here, the anal glands open into crypts and are covered by thin skin flaps, the anal valves. Most anorectal abscesses or fistulae originate here. Above the dentate line, mucosal folds form the columns of Morgagni. The epithelium changes to transitional and thereafter to the columnar epithelium of the rectum. The dentate line also marks the site of a venous portacaval–systemic anastomosis, dividing the internal hemorrhoidal vein plexus above from the external hemorrhoidal vein plexus below. At rest, the anal canal is closed and cannot be visualized adequately with the flexible endoscope. A rigid anoscope or a small speculum gives a much better view.

The external anal sphincter is under voluntary motor control. Strong in its lateral and posterior aspect it is connected with the puborectalis portion of the levator ani muscle. The intersphincteric groove, which can be felt on digital palpation, separates it from the internal anal sphincter. The internal anal sphincter is part of the autonomically innervated circular smooth muscle layer of the gut.

Rectum

The rectum is about 12 cm long from the anus to the rectosigmoid junction. The lower perineal flexure curves posteriorly; the higher sacral flexure curves slightly anteriorly. The lower third of the rectum, the rectal ampulla, lies below the pelvic peritoneum and can become quite dilated. There are three mild lateral curves marked by the valves of Houston: two to the left and the middle and one to the right. It is important to inspect behind these areas carefully, as small lesions are easily overlooked. The middle valve corresponds to the anterior reflection of the peritoneum. The posterior peritoneal reflection reaches only to the rectosigmoid junction. The main muscle layer is longitudinal. The rectum is supported from connective tissue bands to the sacrum and pelvic wall.

Sigmoid

Approximately 16 cm from the anal canal, the rectum merges with the sigmoid colon. Longitudinal muscle fibers form the characteristic taenia sometimes seen during sigmoidoscopy. As the stool in this part of the bowel is formed and sometimes hard, the muscular fibers are strong between the haustral folds. The length of the sigmoid is quite variable, from 40 to 70 cm. The rectosigmoid junction can be quite angulated. The sigmoid bends posteriorly into the hollow of the sacrum, then moves to the right and anteriorly before moving posteriorly again to join the descending colon in the left colon gutter. Overall, it forms a clockwise twist. The sigmoid is surrounded by peritoneum. Pelvic surgery, infection, endometriosis and diverticulitis often lead to significant scarring with fixation of the sigmoid, rendering endoscopic examination difficult or painful.

Descending colon

The descending colon has a short mesentery and runs straight until it bends anteriorly and medially at the splenic flexure to join the transverse colon. If the sigmoid colon mesentery is long, the sharp curve entering the descending colon may be difficult to pass, especially when the bowel is over distended with air. In the descending colon as well as the sigmoid colon, the outlines of the lumen and colon are circular. Rarely, a triangular shape, characteristic for the transverse colon, can be seen in the descending colon. In addition to circular folds, there is also a varied amount of contraction of the circular muscle that causes significant narrowing of the bowel lumen. In about 10–15% of patients, the descending colon has a long mesentery. This renders surgical procedures easy but makes endoscopy difficult.

THE EXAMINATION

Positioning of the patient

Sigmoidoscopy is an embarrassing examination for most patients. A quiet environment and a friendly, concerned staff can help alleviate fear. Some patients feel more comfortable if a sheet covers their abdomen and legs. The traditional position for flexible sigmoidoscopy is left lateral with the hips and the knees flexed. Rarely does this position have to be changed during the examination. The examiner stands on the patient's right side. The assistant should have easy access to instruments, biopsy forceps, or specimen containers. If the main focus of the examination is the sigmoid and descending colon, it may also be adequate to have the patient lie on the right side so that most of the sigmoid is no longer in a dependent position. The examiner then has to move to the patient's left side.

Handling of the endoscope

One-hand vs. two-hands technique

The endoscope is held in the left hand. The web between the thumb and index finger supports the auxiliary cable while the fourth and fifth fingers wrap around the control mechanism housing. The first finger is used to operate the upper suction and lower air/water buttons. The thumb is used to turn the larger up/down angling control wheel. The second finger assists, especially for more extensive movements. It can steady the wheel while the thumb moves on to the next nub on the wheel. The smaller right/left control wheel is more difficult to reach, especially for people with small hands. Often the same tip control effects can be created by using only the up/down control wheel and applying clockwise or counterclockwise torque to the insertion shaft along its length. In this way, the right hand is free to support and control the insertion shaft.

The insertion shaft is most easily manipulated if held between the thumb and first two fingers in the direction of the anus, about 15–20 cm away, so that rarely does one need to reposition the grip. Using gauze to hold the insertion shaft may improve tactile control. As time goes by, one develops a feeling for the force needed to advance the insertion shaft in each situation and the necessary degree of twisting to get around a bend. This assures the best eye/hand coordination and is the smoothest and most efficient way to use the instrument.

People with small hands may find it difficult to perform all control functions with the left hand. One can use the right hand to assist in manipulation of the turn controls, especially the smaller right/left angling wheel. In this case, an assistant may be asked to move or steady the insertion shaft on command. The assistant may also develop a feeling for insertion shaft advancement, but more often pushes either too much or too little. The team approach is usually slower and more awkward, because the subtle clues needed for smooth advancement are difficult to communicate, making coordination difficult. A video monitor is very helpful in this situation.

Basic movements

It is important to familiarize oneself with the endoscope initially, i.e. to get a 'feel' as to how it moves.

I recommend performing the maneuvers summarized below in the open, to observe their effects upon the insertion shaft and tip of the instrument.

Forward Using the single-hand technique, the right hand controls the shaft by advancing it gently. It is very important that this is done cautiously, leaving time for steering and twisting movements. The insertion shaft should be attempted to be kept as straight as possible, because in this position it is the most responsive to control wheel movements. Twisting is only effective on a straightened shaft. Pushing too hard causes the shaft to bend, which may lead to disorientation and loop formation; it also causes pain to the patient and increases the risk of complications. It is not always necessary to have a perfect view of the entire lumen to advance the endoscope, but one should always be certain of the lumen's direction. On rare occasions, the tip can become impacted, allowing only the shaft to move without forward progress. Loss of responsiveness is often due to loop formation, which has to be resolved for the examination to proceed. Feeling the instrument and listening closely to signs of discomfort from the patient are the best methods to achieve a smooth, painless examination.

Pulling back This is the most important movement to regain vision, to determine the direction of the lumen and to straighten the shaft. More time is gained by pulling back than by blindly trying to overcome an obstacle. The best time to inspect the mucosa closely is during withdrawal. Occasionally, a sharp fold or acute bend can quickly slip by; it is very important to ensure that all lumen walls are carefully inspected; sometimes the instrument has to be re-advanced to accomplish this.

Up and down This is done best by turning the larger inner control wheel with the left thumb. Persons with larger hands will also be able to use the middle finger. Excessive angulation leads to a 'walking stick' configuration, which renders further advancement difficult and makes lateral movement inefficient. Steering movements should be made cautiously with close observation of the mucosa, taking into consideration all visual clues to ensure smooth progress. Hectic, excessive movement renders orientation more difficult and prolongs the procedure.

Left and right With the single-hand technique, the smaller, outer left/right control wheel is rarely used. Instead, the instrument tip is flexed upwards and then the shaft is turned right to effect a right turn, or it is turned left to effect a left turn. When the tip is pointing downwards, a right turn is achieved by twisting the shaft to the left and a left turn is achieved by twisting it to the right. With manipulation of the shaft, the angle of the turn can also be adapted. One has to be careful not to bend the tip too much as this decreases the effectiveness of the lateral twist. With practice, this will become one of the most important techniques for smoothly advancing the endoscope, avoiding loop formation as much as possible. Thus, nearly the entire examination can be performed without once using the lateral, right/left steering wheel. Sometimes one can also steady the shaft on the examining table and then use the right hand to manipulate the control knobs. If the examination is performed with an assistant advancing the insertion shaft, then the right hand can be used to manipulate the control wheels. However, if this is done too often, the examination will be prolonged.

Twisting This is a very important maneuver. When the shaft is completely straight, a twisting movement rolls the instrument around its axis, improving the position for suctioning or biopsy or to get into a better position prior to curving around a bend. This movement should feel very easy. When the shaft is straight and the tip bent, twisting causes movement to the right or left, as explained above. Twisting a loop will reduce the size of the loop or resolve it. The sigmoid colon extends in a clockwise spiral through the pelvis to the descending colon. Twisting this loop in a clockwise fashion thus shortens the sigmoid and allows the tip of the instrument to move on.

Torque This means twisting the insertion shaft while advancing or withdrawing. Clockwise torque is particularly helpful in the sigmoid, owing to the anatomy of its course. Rarely, when there is a very long mesentery, counterclockwise torque is better. This decision has to be made empirically each time.

Retroflexed view The interior-proximal portion of the anal canal and base of the rectal vault are difficult to see during insertion or withdrawal of the endoscope. The flexible instrument therefore should be completely turned around, so that this area can be inspected. Most often, the rectum is large enough to permit this maneuver. However, this may not be possible in a small rectum scarred from inflammation or extrinsic compression.

Flexibility The scope straightens itself when it is pulled back by release of the steering controls. Insufflation should be maintained during this process. One should closely watch the mucosa as it slides by for hints as to where the lumen is located. The partial view that is gained is also helpful in finding the lumen again.

Air It is important throughout the examination to use air gently, because too much air will cause discomfort to the patient. By holding the left forefinger slightly above the button, air flow can be closely regulated. If more air is necessary, then the button can be completely occluded. If the patient complains of discomfort or if the colon is completely distended, then the finger can be moved to the suction button to reduce the air. The goal is to get by with as little air as possible to reduce patient discomfort, and to keep the colon soft, pliable and as short as possible.

Suction The suction channel in most instruments exits at the 5:00 o'clock position which should be directed towards the fluid pool. Suctioning under direct vision is most efficient. Otherwise, too much air may be removed which may impair vision even further. Sometimes, a small piece of mucosa gets trapped in the suction channel. Retracting the instrument or insufflating air or water may dislodge it. Rarely, a syringe has to be attached to the suction channel using an intravenous catheter and fluid or air pushed through to clear the channel. After this procedure, the mucosal surface will show small red spots or polypoid projections.

Irrigation Pressing the water button results in cleaning debris from the lens. To clean the colon wall, however, a syringe should be used to irrigate water through the suction/instrumentation port.

Orientation

Determination of distance

Owing to the enormous variability of colon length, as well as the differences induced by colon distension, colonoscopists may be mistaken about the location of a suspicious lesion in 30% of patients or more. This is especially true for partial colonoscopies or sigmoidoscopies. To avoid this problem it is best to make a determination at the time of withdrawal, when there is less distension. During withdrawal, the splenic flexure is at 50 cm, the descending colon at 40 cm and the sigmoid at 30 cm. If more than a 45-cm scope has been inserted at the sigmoid–descending colon junction, a significant loop is probably present. The internal appearance of the colon may be helpful; however, it does not definitely distinguish between the sigmoid and descending colon. In thin patients, the scope may be palpated, especially in the anterior loop of the sigmoid. Transillumination of the abdominal wall is rarely helpful for the same reasons.

Moving around a bend

This maneuver is frequently necessary in the curved sigmoid and has to be accomplished with great patience. Not only does the instrument move, but the sigmoid moves as well, and its bends may change during the examination. It is best to evaluate a bend from a distance and move the scope very slowly, aiming for the darkest area to avoid very sharp turns. If the angle of the scope head is placed correctly, a short pull back flattens out the bend and improves vision. In this way adequate visualization of the lumen is usually obtained. Sometimes, though, it is necessary to pull the scope back further and make a second attempt to get around the bend. Occasionally, it may be necessary to push bluntly for approximately 2 cm. If the tip is pointed in the right direction, then the view should be regained soon. Again, this should be done very carefully to avoid excessive pressure, pain and injury. During this maneuver careful attention must be paid to any patient discomfort. Under no circumstances should the examiner persist in cases of significant pain or mucosal blanching, both signs of excessive pressure. After the bend is passed, the instrument is withdrawn slightly until shaft resistance is felt again and the tip is noted to move back. In this manner, the shaft remains as straight as possible. Use of a mild twist during withdrawal also avoids loop formation and, owing to the anatomical configuration of the sigmoid, often pushes the scope further inside.

Loops

Loops can cause significant discomfort to the patient, owing to stretching of the mesentery. Immediate withdrawal of the scope relieves this pain. Because of the sigmoid anatomy, most loops have a clockwise twist; therefore, it is helpful to twist the shaft clockwise during withdrawal. During this maneuver, bending of the tip slightly lessens back slippage during loop reduction. The amount of scope that needs to be removed depends on the size of the loop. When the tip starts to recede and the shaft transmits sliding movements smoothly to the tip, then the loop is removed. A slight clockwise twist during reinsertion prevents reoccurrence of looping in this area. Ten per cent of patients may have a longer descending mesocolon, and a counterclockwise spiral can form. To reduce it, the shaft has to be twisted counterclockwise during removal. Again the decision regarding which way to turn has to be made by trial and error.

If the sigmoid mesentery is moderately long, an N-loop may form. Here, pain may be caused by stretching of the mesentery. In addition, the angle to enter the descending colon is extremely narrow, making further progress difficult. Reducing this loop often leads to automatic advancement of the shaft into the descending colon.

A very long sigmoid mesentery may twist around its insertion and form an α loop. Again, this may be painful, but in this instance the entrance into the descending colon is very easy.

How to find the lumen

During the course of an examination, the view of the lumen may be lost. If the scope has contact with the mucosa, only red will be seen. In this case, the instrument is pulled back slightly without changing the tip position. Watching the mucosa during this maneuver is helpful as it appears to move towards the lumen. Scope withdrawal, followed by careful reinsertion, is the most important movement to find the bowel lumen. Other means to locate the lumen are as follows:

(1) The wall closest to the scope will appear lighter. Pointing the scope towards the darker areas will help in finding the lumen.

(2) The muscle fibers forming the circular haustral folds can also be used to find the lumen, which is in the center of the arc.

(3) The mucosal vascular patterns generally point towards the lumen or towards the direction of a bend.

(4) Sometimes the taenia coli can be seen as a longitudinal indentation pointing to the bowel lumen.

(5) When the bowel is spastic or deflated, the lumen is in the center at the convergence of the folds.

Sometimes, it can be difficult to find the lumen in patients with diverticular disease. If the scope is inserted into a large diverticulum, one sees a round, central, dark area completely surrounded by lighter mucosa. In contrast, the colonic lumen is characterized by folds extending from the center laterally.

Course of the examination

During sigmoidoscopy, it is important to remain focused at all times to keep the view from slipping away. Both eyes must be kept open and the double pictures not related to the scope must be suppressed. A video monitor renders the examination much more comfortable for the physician and also offers the patient an opportunity to view the examination if so desired. The operator should avoid getting sidetracked during the examination and remember that most orders to the assistant can be given without eye contact.

Prior to the examination, the instruments should be carefully checked to ensure that all systems are operative. The insufflation channel should be free of water and the insufflation pressure adequate.

Prior to insertion of the instrument, the perianal area should be inspected carefully and a rectal examination should be performed to evaluate the lower rectal mucosa and to coat it with a lubricant or local anesthetic. This will dilate the sphincter slightly so that it will better accept the scope. It is best to insert the instrument with the bending section supported by the examiner's finger slightly laterally, especially if there is scaring or tight muscle tone. It may help to start insufflation at this point. Alternatively, a finger can first be inserted and the instrument pushed along this guide. If the sphincter is very relaxed, the instrument can be inserted directly. Sometimes it helps to ask the patient to bear down during this maneuver.

Immediately after insertion of the scope, the mucosa collapses around the scope and a view can be gained only after insufflation. Pulling back with slight rotation usually establishes a view. The scope is then rotated until fluid lies inferiorly and can be suctioned. Care should be taken to submerge the suction port prior to activating it so that air is not removed and visualization lost. Cleaning the rectum avoids spillage of stool during the remainder of the examination. A retroverted view is often necessary for complete evaluation of the lower portion of the rectum. This is achieved by applying complete angulation and slightly moving the instrument forward. In a scarred rectum, this may not be possible, but, in this instance, there is also much less danger of missing a lesion with the straight view.

Advancing the scope through the sigmoid can be difficult, owing to the many folds and curves, especially if the bowel is fixated after previous surgeries or infections. The entire examination is characterized by gentle pushing and retracting as necessary. The initial goal is to advance the scope, with close inspection on the way back. Therefore, time should not be wasted in suctioning all the fluid out, although large amounts of fluid or solid stool should be removed if advancement is impaired. The scope then follows the course of the sigmoid looping anteriorly, then posteriorly. If the sigmoid mesentery is quite short, this may cause some discomfort for the patient. In this situation, loops rarely occur and the passage into the descending colon may be easy following an upwards curve. If the sigmoid mesentery is long, the sigmoid can be distended excessively, creating an N-loop. This leads to a narrow angle at the entrance into the descending colon, which has a shorter mesentery and therefore is much less mobile. By gently advancing the instrument, pulling it back slightly, and using minimal gas, the N-loop can be kept as small as possible, facilitating the instrument's entry into the descending colon. By using a clockwise twist, most of the loops can be straightened out eventually. If the sigmoid mesentery is very long, sometimes it can be twisted so that an α loop is formed. By using the α-loop method, no acute bend between sigmoid and descending colon will occur and the scope can be advanced without difficulty. Sometimes, one can even try to produce an α loop voluntarily because it facilitates further advancement of the scope; however, this is not always possible, especially if the sigmoid mesocolon is short or if there are adhesions. The acute bend between sigmoid and descending colon may be seen at 40–70 cm, depending on how large a loop has been formed. Decreasing the air pressure, moving the instrument back, and then advancing it with a clockwise twisting motion is an important maneuver to reduce the sigmoid loop. The round descending colon is usually passed easily, because it is almost straight.

Inspection is done during withdrawal. Careful evaluation is critical, so that lesions are not missed. Cutting corners at this step decreases the value of the entire examination. If mucosal folds slip by too fast, the scope is simply advanced again and the area re-examined. To enhance the examination all fluid and stool obscuring the surface should be removed, so that lesions are not overlooked. Very small polyps are best seen when the bowel is maximally distended.

Problem areas

Adhesions

Pelvic surgery may cause scarring leading to decreased mobility of the sigmoid. In some cases this does not complicate the examination because loop formation is less likely. After hysterectomy, the anterior wall of the lower sigmoid is usually fixated, but this rarely causes significant deformation. Radiation therapy can also fix the sigmoid in the pelvis. Diverticulosis, especially after several infectious episodes, can also cause significant scarring, mostly in the mid-sigmoid.

Force should never be used, the complaints of the patient should be listened to carefully, and the operator should be patient and prepared to stop. A pediatric scope may make the examination easier. At times referral to a specialist is necessary.

Diverticulosis

Again, the examination in this setting has to be performed with great patience. Insertion should be slow, with care taken not to lose the view. Narrow bends should be anticipated. Hypertrophic circular musculature and occasionally large diverticula, especially when distended with gas, make it quite difficult to determine the lumen.

Severe inflammation

Patients with acute severe inflammatory symptoms or chronic severe inflammatory bowel disease are also at increased risk for complications. One has to be extremely gentle during the examination to avoid injury. Referral to a specialist is often necessary.

Pelvic mass

Large pelvic masses may cause extrinsic compression of the colon. Mostly examination is still possible unless there are significant adhesions. A more difficult problem arises when the bowel wall is involved with tumor. In patients with advanced endometriosis or ovarian cancer, the examination may be very difficult.

When to quit

Sigmoidoscopy is complete when the angulated sigmoid–descending colon junction is reached at 40–60 cm, depending on individual anatomy and colon distension.

Obviously, if solid stool is obstructing the lumen, the examination will be incomplete. Initially additional enemas are used to evacuate the bowel. If unsuccessful it may be necessary to reschedule the procedure. Liquid stool and enema fluid can usually be removed and should not hinder the examination.

Excessive patient discomfort is another reason for aborting the procedure. Scarring with fixation of the bowel wall or intense inflammation may be the cause. In this situation, referral to a specialist or further evaluation by barium enema is recommended.

Similarly, the scope should never be forced through a narrowing. The risk of injury is high in this situation and referral to a specialist is recommended. Sometimes, the use of a pediatric scope is worthwhile in this situation.

Complications

Complications do occur, even if the examiner is careful and skilled. It is therefore very important to inform the patient adequately regarding the advantages and disadvantages of the procedure, as well as the risks and complications[3]. A well-informed patient is also more co-operative, can participate better in the examination, and thus can help prevent problems. Also, the operator should be attentive to the patient to minimize the anxiety of the examination. If the screening examination is embarrassing or very painful, the patient may not agree to a second examination and a valuable screening tool is lost.

In addition, it is important to recognize a high-risk situation early, in hopes of avoiding a complication. Ultimately, it is better to reschedule the examination if preparation was inadequate or refer to a specialist than to risk a complication. Being aware of the possible complications, especially during difficult examinations, helps one to notice a problem early and avoid more severe sequelae. Listening carefully to a patient's complaints and taking them seriously is an important initial step.

Most complications occur in the initial phase of the learning curve[1]. Therefore, novices should be extra diligent. In addition, even after adequate training, it is important to perform the examination on a regular basis to maintain dexterity, judgement and confidence.

There are few publications detailing the overall complication rate of sigmoidoscopy. Although many publications evaluate colonoscopy, most of the complications are located in the rectum and sigmoid. A compilation of several studies evaluating over 40 000 patients revealed a risk of perforation of 0.012% and of bleeding of 0.006%[2]. However, there were no deaths related directly to sigmoidoscopy. Evaluating the number of malpractice claims for gastroenterological procedures, Gerstenberger and Plumeri did not find a significant difference between sigmoidoscopy and more extensive colonoscopy[4]. Diagnostic error, mainly the failure to diagnose a malignancy, and performance error were the two most important factors. Complaints regarding monitoring were less frequent.

Vagal reaction

Distension of the bowel or stretching of the mesentery can create a vagal reaction causing light-headedness, nausea, diaphoreses and low blood pressure. Sometimes these symptoms manifest only after the procedure. Consequently, patients should lie down after the procedure and have their vital signs monitored closely. If any of the above symptoms occur during the examination, the scope should be removed and the patient should expel all gas. The Trendelenburg position may help in restoring blood pressure initially. Rarely are intravenous fluids necessary. Almost all reactions are of short duration. Patients with a history of cardiac problems or arrhythmias, however, are at increased risk of sequelae and should be monitored longer.

Cardiac problems

Cardiac problems occur very rarely during sigmoidoscopy. Monitoring is not generally necessary, except for frail patients with a significant history of cardiac disease or arrhythmia.

Medical complications

Medical problems such as arrhythmia, hypoxia or infections are extremely rare after sigmoidoscopy. Only case reports are available.

Perforation

This is the most feared complication[5,6]. It mostly occurs after the instrument tip is pushed too vigorously or after the shaft is forced into a large loop with excessive lateral pressure on the bowel wall. The latter is probably the most frequent way a laceration of the sigmoid occurs. In this case, the tip of the instrument is in the correct position and the lens transmits a normal picture. Rarely, overdistension with air can also lead to perforation, especially when the ileocecal valve is competent. Here, the only initial sign may be the inability to distend the bowel further. These injuries usually occur in the cecum, even if the scope was not advanced this far, because the cecal wall is very thin while the bowel diameter is quite large. Diverticula can also be the site of pneumatic laceration, again because the wall is very thin and does not have a muscular layer. Rectal perforations are extraperitoneal and can lead to abscess and fistula formation. The clinical signs are initially less significant. Sigmoid perforations are intraperitoneal and lead to free air in the abdominal cavity and peritonitis.

The patient should be closely monitored during each examination. Pain, if felt at all, should be brief and of only moderate intensity, even in patients without premedication. In sedated patients, one has to be even more careful; a short moan could be the equivalent of a scream in an unmedicated patient. The experienced examiner also develops a feel for the maximum force allowed to advance the scope, so that injuries with the tip or with a shaft loop are avoided. Direct perforations are seen immediately when the scope enters the peritoneal cavity. A lateral laceration due to the shaft may not be initially obvious, but it may not be possible to distend the bowel further. Perforation of a diverticulum may be even more difficult to detect. If a perforation of any kind is suspected, the examination should be aborted immediately.

Most perforations are obvious, identified easily during the examination or strongly suspected because of severe abdominal pain, change in vital signs, or signs of peritoneal irritation[6]. The patient should be immediately informed that a perforation has occurred or is suspected. Occasionally the diagnosis is delayed, especially if the perforation is incomplete or retroperitoneal, or the patient's signs and symptoms are blunted owing to previous radiation therapy. Close observation is therefore recommended in case of any suspicion of perforation. Intravenous access should be gained, vital signs should be checked, and the patient should then immediately be transferred to a hospital for observation. Close clinical monitoring should continue, including a basic preoperative laboratory panel and flat and upright abdominal X-rays[7]. A surgical consultation may be necessary. If clinical suspicion is high, broad-spectrum antibiotics with appropriate Gram-negative coverage should be started and the patient should be informed about the possibility of emergency surgery.

An enema with water-soluble contrast may be used to determine the presence of a laceration. Barium is contraindicated in this situation because it can cause a significant peritoneal reaction. A stable patient without peritoneal signs on examination and minimal pneumoperitoneum may be followed conservatively[8]. However, if there is a perforation with peritoneal signs, surgical exploration should not be delayed.

Bleeding

Significant, prolonged bleeding is rare after diagnostic sigmoidoscopy. It is mainly related to mucosal abrasion. Patients with coagulation defects or significant inflammatory changes are at increased risk. As always, the clinician must weigh the benefits of completing the examination versus the increased risk of complications.

In patients who have a biopsy, significant bleeding is rare. If it occurs it is usually seen immediately. After polypectomy, however, bleeding can occur several days later.

Bleeding from the anal canal can usually be stopped by pressure. Sometimes a proctoscope or rigid sigmoidoscope can be used to apply silver nitrate to a small area. In all other instances, referral is recommended for cauterization or injection with epinephrine. Very rarely is angiography or surgical intervention necessary.

DOCUMENTATION

Accurate documentation is always vital. A checklist is the most efficient way to ensure that documentation is complete. In addition, it facilitates computerized record keeping and can be used to generate full reports for the medical record or the referring physician. Prepared endoscopic databases are available commercially, and can be used on personal computers. The list presented here (Table 1) is geared towards the general practitioner, who mostly does screening examinations, and takes only a few minutes to complete. Particular attention should be paid to abnormal findings and their location and appearance described as accurately as possible.

Post-examination instructions

If vital signs are normal after an uncomplicated examination, patients can resume routine activities and a regular diet. The physician should take the time to explain all findings, especially if further evaluation or treatment are necessary. It is best to include detailed post-examination instructions in the general information sheet, so that the patient knows beforehand what to expect and can prepare for it. Some minor abdominal discomfort and cramping, as well as mild rectal spotting, are normal, especially if a biopsy has been performed. The patient, however, should notify the physician immediately if there is severe pain, nausea and vomiting, severe rectal bleeding, or fever. If sedation became necessary, the patient should be advised not to drive a car, drink alcohol or make important decisions until the next day.

TRAINING

Sigmoidoscopy is one of the easier endoscopic procedures. Nevertheless, training is necessary to negotiate the sometimes difficult bends and to establish a reliable diagnosis. It takes practice to optimize eye/hand co-ordination and to develop a feeling for the scope. This may be more difficult for some than for others. It is therefore difficult to set strict standards for the number of examinations necessary to reach competence. In addition, a skill once gained has to be used regularly, so that it is not lost.

Although books, slides, videos and even computerized simulators are important learning tools, nothing can replace apprenticeship with a good teacher. After watching several examinations performed by an expert, the novice can then be phased into doing more and more of the endoscopy with the teacher watching and correcting. The American Society of Gastrointestinal Endoscopy (ASGE) and the American Academy of Family Physicians (AAFP) recommend at least 25 supervised examinations[9]. Schertz and associates[10] suggested a thorough theoretical introduction, as well as 25 supervised examinations to reach competence. Similar recommendations were published in other reports. It is important not only to be able to handle the scope, insert and remove it, but also to be able to determine the diagnosis. This aspect can be improved only with experience. As many as 50 or more examinations may be necessary to see the most frequent pathology. Groveman and colleagues[11] sent questionnaires to 1153 participants in a 1-day workshop on sigmoidoscopy, and received a 66% return rate within 1 year. Overall, more than 17 000 examinations had been performed with a cancer detection rate of 0.9%, which is consistent with the literature. Four severe complications were

THE SIGMOIDOSCOPY EXAMINATION

Table 1 Flexible sigmoidoscopy report

Patient Name	:			
Date of Birth	:		Clinic Number:	
Date	:			
Examination	:	Proctoscopy	Sigmoidoscopy	Biopsy
Indication	:	Screening	Other:	
Pert. History	:			
Symptoms	:	No	Yes:	
Allergies	:	No	Yes:	
Medical Problems	:	No	Yes:	
Patient Instructions Received			Consent Signed	
Abdominal Examination:		Normal	Other:	
Vital Signs	:	T:	R:	P:	BP:
Medications	:	No	Yes:	
Prep	:	good	incomplete but acceptable	poor
Anal Inspection	:	Normal	Other:	
Rectal Examination	:	Normal	Other:	
Images	:	No	Yes	marked with (P)
Biopsy	:	No	Yes	marked with (B)
Extent Visualized	:			
Findings	:	Normal	Other:	
Comments	:			
Complications	:	No	Yes:	
Tolerance	:	Good	Bad:	
Time Started	:		Time Completed:	
Vital Signs Post-examination :		R:	P:	BP:
Diagnosis	:	Normal	Other:	
Medications Given	:	No	Yes:	
Instructions	:	Routine	Other:	
Return to Clinic	:		Next Examination:	
Endoscopist	:			

noted: bleeding necessitating transfusion in two patients and perforation in another two patients. Thorough theoretical knowledge, which can be gained during a weekend course, has to be complemented by practical instruction, for optimal results. This is also the focus of the joint training programs developed by the ASGE and AAFP. Where participation at an organized program is not possible, apprenticeship with a skilled endoscopist is an alternative. Quality assurance is a continuous process. The American College of Physicians Committee in Clinical Privileges regards at least 15 sigmoidoscopies per year as necessary to maintain competence. Guidelines for proctoring prior to granting hospital privileges for sigmoidoscopy are strongly recommended by the American Society for Gastointestinal Endoscopy[12].

CONCLUSIONS

Sigmoidoscopy should not be taken lightly, but it is not difficult if the following requirements are fulfilled:

(1) Adequate knowledge of the anatomy of the lower gastrointestinal tract;

(2) Practice in handling the scope and the basic movements for the examination;

(3) Close attention to every view during the examination and constant reassessment of the position;

(4) Attention to patient comfort or discomfort;

(5) Close observation during and after the examination for possible complications;

(6) Careful documentation;

(7) Adequate initial training and continued practice.

Readers can find further information in a number of other publications[13–16].

REFERENCES

1. Saunders BP, Phillips RKS, Williams CB. Intraoperative measurement of colonic anatomy and attachments with relevance to colonoscopy. *Br J Surg* 1995;82:1491–3

2. Headley JE, Hodge J. *Surgical Anatomy*, 2nd edn. Toronto: BC Decker, 1990

3. Plumeri PA. Informed consent. In Raskin JB, Nord HJ, eds. *Colonoscopy: Principles and Techniques.* New York, NY: Igaku-Shoin, 1995

4. Gerstenberger PD, Plumeri PA. Malpractice claims in gastrointestinal endoscopy: analysis of an insurance industry data base. *Gastrointest Endosc* 1993;39:132–8

5. Greenen JE, Schmidt MG, Wallace CS, Hogan WJ. Major complications of colonoscopy: bleeding and perforation. *Dig Dis Sci* 1975;20:231–5

6. Schwesinger WH, Levine BA, Ramos R. Complications in colonoscopy. *Surg Gynecol Obstet* 1979;148:270–81

7. Kavin H, Sinicrope F, Esker AH. Management of perforation of the colon at colonoscopy. *Am J Gastroenterol* 1992;87:161–7

8. Echer MD, Goldstein M, Hoexter B, *et al.* Benign pneumoperitoneum after fiberoptic colonoscopy. *Gastroenterology* 1977;73:226–320

9. Vennes JA, Ament M, Boyce HW, *et al.* Principles of training in gastrointestinal endoscopy. American Society for Gastrointestinal Endoscopy Standards of Training Committees, 1989–90. *Gastrointest Endosc* 1992;38:743–6

10. Schertz RD, Baskin WN, Frakes JT. Flexible fiberoptic sigmoidoscopy training for primary care physicians: results of a 5-year experience. *Gastrointest Endosc* 1989;35:316–20

11. Groveman HD, Sanowski RA, Klauber MR. Training primary care physicians in flexible sigmoidoscopy – performance evaluation of 17 167 procedures. *West J Med* 1988;148:221–4

12. American Society for Gastrointestinal Endoscopy. Proctoring and hospital endoscopy privileges. *Gastrointest Endosc* 1991;37:666–7

13. Schapiro M, Lehman GA, eds. *Flexible Sigmoidoscopy, Techniques and Utilization.* Baltimore, MD: Williams & Wilkins, 1990

14. Katon RM, Keeffe EM, Melnyk CS, eds. *Flexible Sigmoidoscopy.* Orlando, FL: Grune & Stratton, Harcourt Brace Jovanovich, 1985

15. Cotton PB, Williams CB. *Practical Gastrointestinal Endoscopy*, 4th edn. Boston: Blackwell Science, 1996

16. Raskin JB, Nord HJ. *Colonoscopy: Principles and Techniques.* New York, NY: Igaku-Shoin, 1995

Section II A Review of Cystoscopy

CHAPTER 5

Indications for cystoscopy

Brigitte E. Miller

Cystoscopy is not a screening procedure. For the general practitioner, cystoscopy is sometimes indicated for the evaluation of patients with persistent or atypical urinary tract infections. In addition, the gynecologist may use cystoscopy in the preoperative evaluation of patients with urinary incontinence and for the staging of gynecological malignancies, as well as for the intraoperative evaluation of bladder and ureteral function. This chapter reviews the indications for cystoscopy, including intraoperative cystoscopy.

INDICATIONS FOR CYSTOSCOPY

Atypical urinary tract infection

Cystoscopy is not indicated for the evaluation of simple urinary tract infections, even if recurrent. However, in patients with atypical presentation, persistent infections and suspicion of obstructive problems or urinary calculi, further evaluation with cystoscopy and intravenous pyelography (IVP) is indicated. The suspicion of urinary calculi increases when urea-splitting bacteria are found on repeat culture. Nickel and co-workers[1] found abnormalities in 59% of these patients using cystoscopy and IVP.

Interstitial cystitis

Interstitial cystitis is a chronic inflammatory condition that is ten times more frequent in women than in men. Patients complain of persistent signs of bladder irritation, urinary frequency and pain. On cystoscopy, interstitial cystitis can be diagnosed in the presence of characteristic petechial hemorrhages (glomerulations) and ulcers that extend into the lamina propria (Hunner's ulcers), which are pathognomonic, but present in only 10% of the cases. Although a variety of micro-organisms can often be detected, it is not certain whether these organisms are the cause of the symptoms or a sign of colonization secondary to epithelial cell damage[2].

Incontinence

Cystourethroscopy has been used extensively for many years in the preoperative evaluation of women with urinary incontinence. In comparison to multichannel urodynamic evaluation, the results are less reliable, with a sensitivity of only 60%[3]. Nevertheless, the information gained by cystoscopy is important, as mucosal abnormalities are diagnosed, which can be important for treatment planning. Cundiff and Bent[4] combined urodynamics with cystoscopy and important findings were noted in 19% of the patients. With cystourethroscopy, benign lesions such as diverticula, scarring, stones and other problems can be diagnosed, and malignant or premalignant lesions can be ruled out[5]. All patients with recurrent incontinence after surgical therapy should undergo cystoscopy. Retained sutures may cause detrusor instability, and small fistulas will cause constant leakage.

Diverticula

Urethral diverticula are not infrequent. On complete evaluation, as many as 5% of patients evaluated in a urogynecological clinic were found to have diverticula. Although these are often asymptomatic, diverticula can cause persistent dysuria, frequency, urgency, hematuria, dyspareunia and recurrent urinary tract infections or mild incontinence. Stone formation is also possible. Diverticula most often develop around the openings of the periurethral glands, especially in the middle third of the urethra. Massage of the anterior vaginal wall during urethroscopy often helps to identify the diverticula, as pus or urine can be observed exuding from the orifice.

Hematuria

Hematuria is the most common presenting symptom of bladder cancer. Gross hematuria and persistent microscopic hematuria necessitates further investigation with cystoscopy and IVP to rule out a renal or bladder tumor. Nickel and colleagues[1] diagnosed four transitional cell carcinomas of the bladder among 120 patients with these symptoms that were initially thought to be due to bladder infection. All patients were over 50 years old. Evaluating patients in an integrated hematuria clinic, Paul and associates[6] diagnosed transitional carcinoma in 6% of patients with microscopic hematuria and in 15% of patients with macroscopic hematuria. Infection was the diagnosis in 17% of patients and urolithiasis in 6% of the patients. Following patients within the Kaiser Permanente Care System for many years, Friedman and co-workers[7] noted a much higher incidence of hematuria in patients as long as 5–6 years before the diagnosis of a bladder malignancy was made. Cystoscopy of these patients is not always easy[8]. Papillary tumors are easy to detect; however, flat neoplasias, dysplasia and carcinoma *in situ* may have a very inconspicuous appearance. In case of doubt or when persistent symptoms cannot be explained, referral to a specialist should be made. Urine cytology is inadequate for screening, as well-differentiated tumors are often missed[9].

Bladder stones

Urinary stones can be of metabolic, infectious, or idiopathic origin. Recurrent and persistent infection is the main symptom. Severe pain is much less frequent than with ureteral calculi. Referral to a specialist for medical or endoscopic therapy is suggested.

Fistulas

Cystoscopy is imperative prior to surgery for a vesical fistula. Additional problems such as infection, areas of necrosis, additional lesions, or possible involvement of the ureters can then be excluded.

Staging of pelvic malignancies

Of all gynecological malignancies, cervical cancer is most prone to involve the bladder. Evaluation of the urinary tract should be included in the pretreatment work-up. Early-stage lesions confined to the cervix or incompletely involving the parametrium on palpation, however, rarely involve the bladder. Therefore, cystoscopy is not necessary. There has been a trend during the past decade to decrease the use of cystoscopy for the staging of cervical cancer and confine its use to advanced lesions. Of course, all patients with even minor symptoms related to the bladder need to be evaluated completely. If a bulky tumor is present or fixation is noted to the pelvic wall, cystoscopy should be performed to rule out bladder involvement. Tumor in the bladder mucosa was seen in 20% of the patients with stage III disease on palpation by van Nagell and associates[10]. Urinary cytology is another way to diagnose bladder involvement, but sensitivity is lower, around 56%[11], with a specificity of 93%[12].

Endometrial and ovarian cancer rarely involve the bladder. Evaluation should be individualized and is necessary only in symptomatic patients.

Follow-up after bladder tumors

As the treatment in these cases is mostly carried out by a specialist, so are the follow-up examinations. It takes considerable experience to detect early invasive or *in situ* lesions.

Preoperative cystoscopy

Preoperative cystoscopy is indicated when involvement of the bladder with tumor or endometriosis is suspected. Ureteral stents can be placed at that time, as advocated by some authors. Overall, stent placement is of borderline benefit in reducing intraoperative ureteral injury. By no means can it replace good surgical technique.

Intraoperative cystoscopy

During urogynecological procedures or oncological surgeries, the urinary tract, bladder and ureters are at risk for injury. Many of these injuries are never detected clinically, but can lead to severe compromise of renal function or to fistula formation. The fistula rate is reported as 0.02–0.35%. With intraoperative cystoscopy, Harris and co-workers[13] noted a complication rate of 4% among 224 patients undergoing a Burch procedure or culdoplasty. Most injuries were related to sutures and could be corrected by removal of the offending suture without further sequelae. Only one additional operative procedure became necessary. They recommended intraoperative surveillance cystoscopy for all these procedures. Pettit and Petrou[14] evaluated patients after vaginal surgery including urogynecological procedures, anterior repairs and culdoplasty. Among 236 patients, seven injuries were detected by cystoscopy and indigocarmine injection. No complications due to this procedure occurred. In conclusion, one could recommend cystoscopic evaluation after all urogynecological procedures. Although the incidence of ureteral obstruction was only 0.33% in a large study by Stanhope and colleagues[15], who evaluated 5379 patients after major pelvic surgery for benign conditions, cystoscopy should not be delayed if there is any suspicion about the integrity of the bladder and ureter.

Transurethral cystoscopy requires special positioning of the patient and additional preparation. To avoid this inconvenience, Timmons and Addison[16] recommended a suprapubic approach. This can be done easily at the operating table. Ureteral catheters can also be inserted if necessary. The small bladder incision heals quickly and completely and no special postoperative care is necessary.

CONTRAINDICATIONS

There are few contraindications to cystoscopy. An elective examination should not be done during an acute urinary tract infection. The patient should be willing and able to cooperate. If a difficult examination is anticipated, such as in the case of significant bleeding, or if severe pain is anticipated, as during evaluation for interstitial cystitis, referral to a specialist and examination under anesthesia is recommended.

CONCLUSIONS

The primary indications for cystoscopy in general or gynecological practice include the following.

(1) Atypical and/or persistent infection;

(2) Gross or persistent microscopic hematuria;

(3) Evaluation for urinary incontinence;

(4) Evaluation of urinary fistulas;

(5) Staging of advanced pelvic malignancies;

(6) Intraoperative assessment of suspected bladder or urethral injury.

REFERENCES

1. Nickel JC, Wilson J, Morales A, Heaton J. Value of urologic investigation in a targeted group of women with recurrent urinary tract infections. *Can Urol Assoc* 1991;34:591–5

2. Keay S, Schwalbe RS, Trifillis AL, Lovchik JC, Jacobs S, Warren JW. A prospective study of microorganisms in urine and bladder biopsies from interstitial cystitis patients and controls. *Urology* 1995;45:223–9

3. Scotti RJ, Ostergard DR, Guillaume AA, Kohatsu KE. Predictive value of urethroscopy as compared to urodynamics in the diagnosis of genuine stress incontinence. *J Reprod Med* 1990;35:772–6

4. Cundiff GW, Bent AE. The contribution of urethrocystoscopy to evaluation of lower urinary tract dysfunction in women. *Int Urogynecol J* 1996;7:307–11

5. Summitt RL. Investigative techniques, assessment of incontinence, and urodynamics. *Curr Opin Obstet Gynecol* 1992;4:548–53

6. Paul AB, Collie DA, Wild SR, Chisholm GD. An integrated hematuria clinic. *Br J Clin Pract* 1993;47:127–30

7. Friedman GD, Carroll PR, Cattolica EV, Hiatt RA. Can hematuria be a predictor as well as a symptom or sign of bladder cancer? *Cancer Epidemiol Biomarkers Prev* 1996;5:993–6

8. Winkler HA, Sand PK. The evaluation and management of hematuria in women. *Int Urogynecol J* 1997;8:156–60

9. Maier U, Simak R, Neuhold N. The clinical value of urinary cytology: 12 years of experience with 615 patients. *J Clin Pathol* 1995;48:314–17

10. van Nagell JR, Sprague AD, Roddick JW. The effect of intravenous pyelography and cystoscopy on the staging of cervical cancer. *Gynecol Oncol* 1975;3:87–91

11. Russell AN, Shingleton HM, Jones WB, et al. Diagnostic assessments in patients with invasive cancer of the cervix: a national patterns of care study of the American College of Surgeons. *Gynecol Oncol* 1996;63:159–65

12. Omigbodun AO, Thomas JO, Adewole IF. Cytologic detection of urinary bladder involvement in cervical cancer. *Int J Gynecol Cancer* 1994;4:401–3

13. Harris RL, Cundiff GW, Theofrastous JP, Yoon H, Bump RC, Addison WA. The value of intraoperative cystoscopy in urogynecologic and reconstructive pelvic surgery. *Am J Obstet Gynecol* 1997;177:1367–71

14. Pettit PD, Petrou SP. The value of cystoscopy in major vaginal surgery. *Obstet Gynecol* 1994;84:318–20

15. Stanhope CR, Wilson TO, Utz WJ, Smith LY, O'Brien PC. Suture entrapment and secondary ureteral obstruction. *Am J Obstet Gynecol* 1991;164:1513–19

16. Timmons MC, Addison WA. Suprapubic teloscopy: extraperitoneal intraoperative technique to demonstrate ureteral patency. *Obstet Gynecol* 1990;75:137–9

CHAPTER 6

Equipment for cystoscopy

Greg Portera

Diagnostic office cystoscopy can be performed adequately with a few instruments, which will be described below. Some of the components are also useful for other procedures, such as hysteroscopy. A light source or monitor is necessary for a variety of procedures. When all new equipment is purchased for the office, care should be taken to ensure that all individual components can be used together. More sophisticated equipment is used for complete urodynamic evaluation, but this is beyond the scope of this book.

HISTORY

There have been many attempts to illuminate and visualize the bladder. The first was by Bozzini in 1806, when he presented his Lichtleiter to the Academy of Medicine in Vienna. The instrument consisted of a urethral speculum and a candle for a light source. However, in 1877, Max Nitze, utilizing Edison's incandescent bulb for a light source, developed an endoscope that is the foundation for the modern-day cystoscope. The first American cystoscope was developed in 1900 by Otis and Wappler, and in 1970 Brown and Buerger invented what has become the primary cystoscope of contemporary use. Wappler continued to improve and refine existing cystoscopic instruments and, working together with McCarthy, devised the Foroblique or McCarthy Panendoscope. The utilization of fiberoptics was the next major advancement and was first developed in 1929. However, it was not until 1954 that Hopkins incorporated fiberoptics into the cystoscope, which utilized an external high-intensity lamp to transmit light through small microfibers of the scope. Hopkins was also responsible for the rod-lens system of the cystoscope which enabled greater transmission of light. In 1973, Tsuchida introduced the first flexible cystoscope with the ability to bend at the distal tip to allow observation of the bladder neck. It was not until several years later, however, that the flexible cystoscope gained its popularity.

CYSTOSCOPES

Rigid cystourethroscope

After the development of the fiberoptic light source and improved optics by Hopkins, endoscopic instruments were developed that combined both urethroscopic and cystoscopic telescopes. All cystourethroscopes have basically the same design, consisting of an outer sheath, obturator and interchangeable telescopes[1]. A bridge may be attached, to allow passage of ureteral catheters, biopsy forceps and other instruments. The Wappler cystourethroscope has become very popular with its interchangeable microlens system to visualize the urethra and bladder.

Cystourethroscopic instruments are measured by their diameter, using the French scale[2]. The diameter, in millimeters, multiplied three-fold, gives the French size. The sheaths come in many different sizes, and the conventional fenestrated sheath used for diagnostic procedures of the bladder is usually 16–21 Fr. However, when operative procedures are performed through the cystourethroscope, 21–24-Fr sheaths are needed to allow passage of different instruments. There are different types of sheath for different procedures. We recommend a short beak sheath with a 12° telescope when visualizing both the proximal and the distal urethra. The outer sheath has one or more ports to introduce distension media, to evacuate fluid from the bladder, or to insert different operative instruments. A fenestrated sheath can be used when the main focus is on evaluation of the bladder. However, it is inadequate when viewing the urethra, because the liquid media will escape the urethral meatus during withdrawal of the cystourethroscope. The walls of the urethra will then collapse around the end of the scope and adequate visualization become impossible.

Obturators can be placed into the sheath to provide a blunt, smooth tip for insertion in the urethra. Different types are available to fit the particular sheath. The closed obturator provides a solid, intact tip for introduction of the instrument into the urethra. The visual obturator accepts a telescope and fills the space between the telescope and the sheath to provide a smooth surface against the urethra and allow visualization during insertion. Often, insertion of the instrument without an obturator is possible if it is done very gently.

Cystourethroscopes now have variable lens systems. These include a straightforward telescope, a Foroblique telescope, or angled telescopes ranging from 0° to 120°. The 0° or 12° (Foroblique) lens is primarily used to look straight ahead to visualize the urethra and the bladder neck. The 30° lens allows visualization of the bladder base and posterior wall of the bladder. This is the most versatile instrument. The 70° or 90° lens is used to inspect the dome of the bladder and the anterior-lateral walls. The 110° or 120° lens (retroview lens) is less commonly used, but permits optimal visualization of the anterior bladder neck.

Flexible cystourethroscope

Although flexible cystoscopy was first reported in 1973, it was more than 10 years later before it became popularized in the USA.

There are several advantages and disadvantages of flexible cystourethroscopes[3]. The chief advantage is that the flexible telescope is better tolerated with less discomfort than the rigid telescope, especially in male patients. Second, the flexible cystourethroscope allows visualization of the entire bladder, especially difficult areas such as the bladder neck and diverticula[4]. This obviates the need for multiple insertions of varying rigid scopes which is often necessary for complete evaluation. In addition, the flexible cystoscope is ideal for patients with scarred or fixed bladders due to previous surgeries or radiation treatments[5,6]. Last, these scopes can be used in the patient who cannot tolerate the dorsolithotomy position. Difficult urethral catheterization is a definite indication for flexible cystoscopy. Under direct vision, a guidewire can be passed, followed by dilators and lastly the catheter. Although this can be done at the bedside, these procedures should remain the domain of the specialist, as extensive experience is needed in these difficult situations.

Disadvantages of the flexible cystourethroscope include its lack of clarity and its small viewing size compared to the rigid cystoscopes. The light source of the flexible cystourethroscope is also inferior, owing to the fiberoptics that also create problems with regard to durability. The flexible cystourethroscope is adequate to evaluate minor hematuria; however, if there is a large amount of bleeding, clot formation, or other debris, this can obstruct the scope. Also, due to the caliber of most flexible cystoscopes (5–7 mm), it is sometimes difficult to acquire adequate tissue with the small biopsy forceps. To ensure that the sample is sufficient, multiple biopsies are often necessary.

Finally, flexible cystourethroscopes are more expensive than rigid equipment. For practitioners evaluating primarily female patients, the advantages of the flexible instrument may not offset the added cost. Ultimately, flexible cystoscopes are not likely to replace rigid scopes completely, but they do have unique capabilities that are a useful addition to a physician's armamentarium.

Monitors

In the past, just the telescope was used for observation, sometimes putting the observer in uncomfortable positions as well. A video system renders the examination much easier. It also gives the patient an opportunity to watch the examination and learn more about the disease process.

Biopsy instruments

Only very small biopsy instruments should be used in the office, so only a small sheath is needed for insertion, causing less discomfort and pain. Tiny flexible forceps can be passed through a rigid scope, and even smaller ones through a flexible scope. Several biopsies may become necessary, but as they can be taken under direct vision, a diagnosis is mostly possible. The advantage of these instruments is that the scope does not have to be moved during the procedure. Larger forceps are available, attached to the sheath. The angle between biopsy forceps and scope is fixed, sometimes making it more difficult to reach a lesion. In addition, the entire sheath has to be removed after the biopsy has been taken, and has to be reinserted if a second biopsy is necessary. This may cause considerable discomfort to the patient. If extensive biopsies are necessary, the procedure should be performed in the surgical suite, so that larger sheaths can be used after adequate anesthesia is achieved. In addition, back-up instrumentation should be available in case large-volume irrigation and cautery are needed.

Light sources

The fiberoptic light system allows transmission of high-intensity light from outside the body to the cystourethroscope. The fiberoptic system is composed of a bundle containing very small glass fibers. Light that enters these fibers is trapped and transmitted down the fibers. In the cystourethroscope, a cord containing fiberoptic bundles is attached to fiberoptic bundles located in the telescope, creating one continuous system. The rod-lens is responsible for transmitting images while the fiberoptic bundle transmits light. Flexible cystoscopes also use fiberoptic bundles from the light source, but they have a separate fiberoptic bundle within the scope used for imaging.

Improvements continue to be made with the intensity of external light sources transmitted through fiberoptic bundles. Currently, the xenon light source is the most efficient lamp in use.

MAINTENANCE AND STERILIZATION OF EQUIPMENT

Cystoscopes and their numerous attachments are expensive instruments that may be used several times a day. It is critical to maintain these instruments with careful cleaning and sterilization to avoid transmission of infectious diseases and to prolong their function[7].

After the examination, the cystoscope should be removed from its outer sheath. All cystoscopy instruments are cleaned by first removing all adherent debris, including blood, tissue, or accumulated irrigation media, with a soft brush or gauze. The instruments should then be thoroughly rinsed with water before sterilization.

After cleaning the instrument, care must be taken to avoid transmission of infectious disease. This is accomplished with sterilization or high-level disinfection. Sterilization procedures destroy all microbial life within the system, including highly resistant bacterial spores. Since instruments may be damaged by the high temperatures achieved in steam autoclaves, cystoscopy instruments must be sterilized chemically through disinfection. In general, disinfection eliminates most pathogenic micro-organisms, but not necessarily all microbes or spores. Disinfection techniques are therefore classified into three levels (high, intermediate and low) based on their antimicrobial activity. High-level disinfection achieves complete chemical sterilization because it possesses activity against spores, fungi and viruses. Available agents include 2% glutaraldehyde (Cidex), 8% formaldehyde solution in 70% alcohol, 6–10% stabilized hydrogen peroxide and ethylene oxide gas.

Cystoscopes should undergo both cleaning and sterilization. Damage to the cystoscope, particularly to the flexible cystoscopes, must be considered in choosing a regimen for disinfection. Two of the most commonly used agents for high-level disinfection are 2% glutaradehyde (Cidex) and ethylene oxide. We recommend submerging the instruments in Cidex for 15–20 min before each case. Cidex must be carefully washed off the instrument after sterilization because Cidex can be extremely irritating to the skin and mucosa.

EXAMINATION ROOM

Any procedure room can be used for cystoscopy. A gynecological examination table is necessary, preferably with full support for the knees and automatic positioning. For short procedures without bladder irrigation, a special drainage area is not required. An intravenous pole or ceiling hook is used to elevate the distension fluid for better infusion. A Mayo stand makes the instruments easy to display and to reach.

CONCLUSIONS

The following items are necessary for basic office cystoscopy:

(1) Gynecological examination table, intravenous pole, connection tubes;

(2) Cystourethroscope with outer sheath, obturator and telescopes (Foroblique, at least 30° and 70°);

(3) Light source and fiberoptic cables;

(4) Container for Cidex sterilization and defined cleaning area.

REFERENCES

1. Robertson JR. Gynecologic urethroscopy. *Am J Obstet Gynecol* 1973;115:986–90

2. Greene LF, Khan AV. Cystourethroscopy in the female. *Urology* 1977;10:461–2

3. Figuerua TE, Thomas R, Moon TD. Taking the pain out of cystoscopy: a comparison of rigid with flexible instruments. *J Louisiana St Med Soc* 1987;139:26–8

4. Tsuchida S, Sugawara H. A new flexible fibercystoscope for visualization of the bladder neck. *J Urol* 1973;109:830–1

5. Kavoussi LR, Clayman RV. Office flexible cystoscopy. *Urol Clin N Am* 1988;15:601–8

6. Fowler CG, Badenoch DF, Thaker DR. Practical experience with flexible fibrescope cystoscopy in outpatients. *Br J Urol* 1984;56:618–21

7. Fozerd JB, Green DF, Harrison GS, Smith PH, Zolite N. Asepsis and outpatient cystoscopy. *Br J Urol* 1983;55:680–3

CHAPTER 7

Patient preparation for cystoscopy

Brigitte E. Miller

No major preparation is necessary for cystoscopy, although the general medical status of the patient and the indications for the procedure need to be carefully evaluated. Complete patient education about the course of the examination is critical, because a well-oriented patient can be more co-operative, making the examination easier to endure and easier to perform. In most cases, the procedure can be accomplished within a few minutes in the office.

MEDICAL HISTORY AND PHYSICAL EXAMINATION

A careful history detailing the patient's symptoms should be obtained, to confirm the indication. Attention to hematuria, recurrent bladder infections, or signs of incontinence or obstruction is particularly important. Chronic infections cause urinary frequency, urgency and dysuria and often a low-grade fever or hematuria. Signs of a neurogenic bladder include persistent urgency or urge incontinence, nocturia, small bladder capacity and poor bladder sensation. Hesitancy, slow stream and the feeling of incomplete voiding may be related to a hypotonic bladder or outlet obstruction. Chronic infections, interstitial cystitis, bladder tumors, or radiation changes often cause chronic pain, frequency, small bladder capacity and hematuria. Stress incontinence becomes manifest during situations of increased abdominal pressure, such as coughing or straining. The occurrence of urge incontinence due to a detrusor problem is unrelated to stress. A fistula manifests itself by intermittent or total urinary incontinence. Recurrent infections and post-void dribbling can be related to a urethral diverticulum. A focused review of systems should be carried out, with attention to previous urinary complaints or procedures, including recurrent infections, stone formation, or pelvic radiation. General medical problems and medications are noted. An inclusive history is often a major step in establishing the correct diagnosis.

The examination includes evaluation of the costovertebral angles and palpation of the kidneys. Careful examination of the lower abdomen and suprapubic area is as follows. First, an enlarged bladder can be felt easily, and tenderness may suggest infection or tumor; second, a pelvic examination should complete the evaluation with special attention to signs of pelvic floor relaxation, masses, or tenderness, especially in the anterior vaginal wall. Notation of cervical lesions, an enlarged uterus, or other pelvic masses should be made.

SPECIAL CONSIDERATIONS

The very anxious patient

Most patients will be able to tolerate cystoscopy without premedication. Detailed explanations about the procedure, a calm environment and a kind staff are the most important factors.

In the rare uncooperative patient, conscious sedation can be obtained, as discussed in Chapter 4 about sigmoidoscopy. With the patient sedated, one has to watch the fluid volume infused into the bladder carefully to prevent overdistension, which could cause injuries.

The elderly patient

Overall, elderly patients are at increased risk for bladder malignancies. Elderly women with atrophy may have narrowing of the urethra leading to an increased risk of discomfort and injury. The smallest possible instrument or a flexible cystoscope may render the examination easier. Local medication with estrogen cream is also helpful.

The pregnant patient

Rarely is cystoscopy necessary in pregnancy. Except for a slightly increased risk of infection, there are no unique concerns.

Table 1 Patient instructions for cystoscopy

During cystoscopy, a small tube will be inserted into your bladder allowing your physician to examine the urethra and bladder wall, check the ureters (connection between kidney and bladder), and take a small tissue sample if necessary.

Preparation

No special preparation is necessary at home prior to the examination. Eat a light breakfast. Take all your usual medications.

Examination

In the office, your doctor or nurse will again explain the procedure to you. Feel free to ask questions. Let us know about new complaints or problems, difficulty moving your hips, or an iodine allergy. Your vital signs will be taken. You will then be placed on the examination table with your legs elevated as comfortably as possible. The outside of your urethra may be cleaned to reduce the risk of infections. A local anesthetic will then be inserted into the urethra. Drapes may be placed over your legs and pelvis to reduce the risk of infection. The cystoscope will then be inserted. This may cause some stinging discomfort. Sometimes a urine sample is taken at this time. Fluid is then inserted into the bladder for better visualization. Please let us know when you feel the bladder is full. The cystoscope is then moved around so that the entire bladder wall can be inspected. Again, let us know if you feel any discomfort. Sometimes an injection of blue dye is necessary to evaluate the ureters. At the end of the procedure, the fluid is drained and the instrument is slowly removed so that the urethra can be examined.

After the examination

You may experience some mild discomfort passing urine after your examination. However, this should resolve soon. You may be offered a medication called Pyridium to help with the discomfort. Sometimes a small amount of blood will be present in the urine, especially if a biopsy has been taken. Please let us know if you pass blood clots, if urination remains painful, or if you develop fever or back pain.

Complications

Cystoscopy is a very safe procedure. Sometimes a bladder infection can develop. Very rarely, the bladder wall can be injured so that surgery may become necessary.

Problems

At any time if you have more questions, feel free to call...

In case of emergency call..

Prevention of bacterial endocarditis

Antibiotics are unnecessary unless the examination is performed in the presence of a significant urinary tract infection or in patients at high risk for endocarditis (discussed previously). Prophylaxis with amoxicillin 3 g orally prior to the procedure and 1.5 g 6 h later can be given to patients at low risk. Gentamycin 1.5 mg/kg intravenously (not to exceed 80 mg) and ampicillin 2 g intravenously prior to the procedure and repeated 8 h later is appropriate for patients at high risk. Vancomycin is suitable for penicillin-allergic patients.

The anticoagulated patient

The risk of bleeding is low during cystoscopy. Coagulation parameters, however, should be checked in patients currently taking anticoagulant medication. If the examination is performed in the presence of significant hematuria due to infection, tumor, or radiation injury, discontinuation of anticoagulant medication is recommended.

The diabetic patient

No diet restrictions are necessary prior to routine cystoscopy. However, these patients are at increased risk of urinary tract infection, and antibiotic prophylaxis is strongly recommended.

PREPARATION PRIOR TO THE EXAMINATION

In contrast to sigmoidoscopy, no special preparations are necessary prior to the examination. Patients

Table 2 Consent form

Name: _____ Date: _____ Time: _____ am/pm

(1) I authorize Dr. _____ to perform a cystoscopy with biopsies of the bladder wall if indicated.

(2) The purpose of the examination has been explained to me and is evaluation for _____
_____.

(3) The risks involved and possible complications have been explained to me, including but not limited to bladder infection and injury to the bladder and urethra, possible surgery and other medical problems. No guarantee or assurance has been given by anyone as to the results that may be obtained.

(4) Alternative methods have been explained to me.

(5) I give my consent to treatments and medical procedures on my body by all qualified personnel working under the supervision of the aforementioned doctor, before, during, and after the procedure to be performed.

(6) I consent to the disposal of any tissue which may be removed.

Signed: _____ Relationship: _____

I have given to the above individual an explanation of the contemplated procedure and its anticipated benefits, risks and potential complications.

Signed: _____, M.D. Date: _____ Time: _____ am/pm

I certify that I was present at the time the above explanation was given and, in my opinion, the subject understands the factors of this consent. I also witnessed the signature of both parties.

Signed: _____ Position: _____

Date: _____ Time: _____ am/pm

---

scheduled for office cystoscopy can continue taking their regular meals and medications, unless conscious sedation is planned.

Antibiotic prophylaxis

Bacteriuria has been described after cystoscopy in as few as 2.8% and as many as 16% of patients, and has not always led to bladder infection. Prophylactic antibiotic therapy decreases this risk significantly. The opinion in the literature is varied. We use a single dose of Bactrim® or ampicillin.

Patient instructions

As always, a well-informed patient is less anxious and more cooperative. Every step of the examination should be carefully explained. Written information is also helpful in reducing the risk of problems or complications. Sample instructions and a consent form are shown in Tables 1 and 2.

CONCLUSIONS

Adequate patient preparation for cystoscopy includes:

(1) Review of medical history and medications;

(2) Physical examination;

(3) Explanations about the course of the examination, including possible discomfort;

(4) Counseling regarding indications, risks and complications;

(5) Consent form.

REFERENCES

1. Montella JM, Ostergard DR. Office urethroscopy. In Buchsbaum HJ, Schmidt JD, eds. *Gynecologic and Obstetric Urology*, 3rd edn. Philadelphia, PA: WB Saunders, Harcourt Brace Jovanovich, 1993:91–6

2. Lawson RK, Taylor AJ. The urologic examination. In Buchsbaum HJ, Schmidt JD, eds. *Gynecologic and Obstetric Urology*, 3rd edn. Philadelphia, PA: WB Saunders, Harcourt Brace Jovanovich, 1993:77–90

CHAPTER 8

The cystoscopy examination in the female

Brigitte E. Miller and Ward A. Katsanis

The female bladder lends itself to easy examination. The smooth surface can mostly be well visualized. On the other hand, it can be quite difficult to detect early or preinvasive malignancy. In this chapter, anatomy and the course of the examination are discussed. It is important that cystoscopy is fast and complete, as the examination can be quite uncomfortable for the patient.

ANATOMY

The female external urethral orifice is a 5-mm vertical slit between the clitoris and vaginal orifice. The female urethra extends upwards over 3–4 cm above the anterior vaginal wall to the bladder neck. Surrounded by extraperitoneal tissues, the urethra is a smooth tube of skeletal muscle positioned close to the vaginal wall. It is firmly supported beneath the symphysis pubis by the pubourethral ligaments. It passes the fascia of the levator ani muscle at the bladder neck. The urethral blood supply is mainly derived from the vaginal arteries, while venous and lymphatic drainage empties primarily into the pelvic vessels. The hypogastric nerves of the pelvic plexus provide smooth muscle innervation and branches of the pudendal nerve innervate the striated muscle fibers of the external urethral sphincter.

The urethral submucosa contains a vascular plexus that produces mucosal coaptation similar to a cushion. It extends throughout the entire course, but is thinner close to the meatus. It accounts for about one-third of the intraurethral pressure. The urethral epithelium is responsive to estrogens. In case of atrophy, local medication with estrogen cream is helpful. The mid- and proximal urethra are both important for the continence mechanism. The highest resting pressure is seen in the mid-urethra due to surrounding fibers of striated muscle. A portion of the proximal urethra is positioned intra-abdominally and is an important part of the continence mechanism.

The empty bladder is situated entirely within the pelvis. However, when filled it extends into the lower abdominal cavity and is covered superiorly by peritoneum. Anteriorly, the space of Retzius is located below the symphysis pubis and the rectus muscle. Posteriorly, the bladder lies on top of the uterine cervix and upper vagina. The inferior and lateral surfaces are situated above the levator ani and obturator internus muscles. The bladder neck is directly above the urogenital diaphragm. Fixation of the bladder is achieved by thickenings of the endopelvic fascia, the medial and lateral pubovesical ligaments, the lateral true ligaments of the bladder, the medial umbilical ligament and the remnant of the urachus. The blood supply originates from the hypogastric system mainly through three vesicle arteries: superior, middle and inferior. In addition, the trigonum area receives blood supply from the uterine artery, and vaginal arteries supply the bladder neck. The hypogastric vein provides the predominant venous drainage and the lymphatics also drain through the pelvis. The hypogastric sympathetic plexus and the pelvic splanchnics provide the nerve supply to the bladder. These fibers mainly accompany the vessels.

The lumen of the empty bladder is similar to a triangle. Filled, the bladder is almost spherical. The bladder wall has three layers of musculature – inner longitudinal, middle circular and outer longitudinal – which are best defined in the trigonum area. The inner layer continues through the bladder neck to the longitudinal fibers of the urethra. In most of the bladder wall, these layers are difficult to distinguish, as their thickness varies and fibers run in many different directions. The mucosa, therefore, has mesh-like folds in the non-distended state, although the trigonum area remains smooth. The bladder epithelium is transitional.

The intravesical portion of the ureters is about 2 cm long. The ureteral orifices are found on the interureteric ridge, about 2.5 cm apart in an empty bladder. The appearance can be variable, usually slit-like, sometimes rounded, occasionally surrounded by slightly raised epithelium.

THE EXAMINATION

Position of the patient

Although the proper position for the examination is quite embarrassing, one should try to make it as comfortable as possible. Any gynecological examination table is appropriate, although a table with knee support is preferable. A small pillow should support the head. Patients with severe cardiac problems will also appreciate mild elevation of the thorax. The buttocks should be placed at the end of the examination table with the legs supported in well-padded knee crutches or stirrups. These should be far enough apart to allow access to the perineum. In patients with hip problems, this may not be possible and assistants may have to support the patient's legs. It is important to avoid excessive abduction in these cases. At no time should the legs be forced into position, especially in the medicated patient.

Preparation of the patient

During the procedure, bacteria from the skin can be introduced into the bladder, leading to infection. Antiseptic preparation of the entire genital area including inner thigh and mons pubis will decrease the germ count and the risk of bladder infection. Shaving is not necessary for diagnostic cystoscopy. Initially, the area should be washed with antiseptic soap and warm water followed by painting with an antiseptic solution such as Betadine®, or quarternary amine solutions in iodine-allergic patients. Ongoing dialogue during the examination will keep the patient informed and reduce embarrassment and anxiety.

Adequate draping decreases the risk of infection and may render the examination less embarrassing for the patient. Single drapes can be used to cover the legs, lower abdomen and rectum. However, it is easier to use a special disposable cystoscopy drape that covers these areas in one piece and also includes a screen for the drainage pan.

Some authors have recommended a technique without draping or cleansing of the vulva. With this technique, it is important to separate the labia as much as possible. Sterile gloves may not be necessary if one avoids touching the cystoscope in the area to be introduced into the bladder. The infection rate with this technique does not seem to be higher.

With a small cotton-tip applicator or catheter tip syringe, a local anesthetic agent (2% lidocaine) is inserted into the female urethra. This should be done 5–10 min before the examination for maximal effect. A viscous medium is preferred as it remains in the urethra and does not flow easily into the bladder. If careful inspection of the urethral mucosa is important, one should avoid local anesthetics as they can cause a slight reddening of the urethral mucosa.

For short procedures such as stent removal or small bladder biopsy, intravesical local anesthesia or a bladder block can be given. For local anesthesia, the bladder is drained with a catheter and then 50 ml of a 4% lidocaine solution is instilled and left in place for at least 5 min. Once this fluid is drained, the examination can begin. For the bladder block, the bladder pillars are located at 10:00 and 2:00 o'clock at the anterior vaginal fornix above and slightly lateral to the cervix. Five milliliters of 1% lidocaine solution injected here also decreases bladder pain during manipulation. If the cervix is absent, the medication is injected below the vaginal wall in the same area.

If significant manipulation is planned, the examination is best performed under anesthesia or conscious sedation, since a larger, more uncomfortable operative sheath will be inserted. This is also true in cases of significant urethral stenosis, bladder bleeding, or evaluation for interstitial cystitis[1].

Physician preparation

Sterile gloves are recommended to ensure antiseptic technique. A water-repellent sterile gown may be used, especially if a prolonged procedure with irrigation is expected. This is unnecessary for a short diagnostic cystoscopy.

Distension media

A variety of distension media can be used for cystourethroscopy depending on the type of procedure to be performed. Liquid media are most frequently used and can be divided into conductive media (normal saline, lactated Ringer's) and non-conductive media (sterile water, glycine). All distension media should be infused at body temperature. It is also prudent to limit the height from which the medium is infusing to 60 cm above the bladder. This will limit the distension pressure to 60 cmH_2O or less and minimize patient discomfort. In the sedated patient, one must carefully monitor the volume of fluid infused to avoid overdistension and possible bladder injury. Gases such as carbon dioxide or air can also be used.

Sterile water is an excellent medium for short, diagnostic cystourethroscopy. It does not mix well with blood, which improves visualization in cases of hematuria. If the procedure is prolonged and large amounts of fluid are necessary, there is a risk of increased absorption, leading to electrolyte imbalance or hemolysis. If the intent is to observe ureteral function, D10 is another good option. The excreted urine can be seen quite well in this distension medium. When indigocarmine is used to stain the urine, it does not mix and thus does not obstruct vision. After one ureteral orifice has excreted the blue solution, the other orifice can still be observed clearly.

Normal saline and lactated Ringer's can be used in larger volumes without the risk of electrolyte imbalance or hemolysis. Unfortunately, these media mix well with blood, which obscures vision in cases of bleeding. These media cannot be used in conjunction with electrical devices, as they are conductive and can cause injury to the patient and the operator.

Glycine is the most versatile medium for urological procedures. Relatively large volumes can be used without causing electrolyte imbalance or pulmonary edema. Glycine does not mix well with blood and, since it is not conductive, it can be used with electrocautery.

Air or carbon dioxide is used mainly during flexible cystoscopy. Bladder distension is comparable to that with other media. The bladder has to be totally empty for complete evaluation. In cases of bleeding, visualization is often better. Patients seem to tolerate this method as well as or even better than distension with fluid.

Handling the scope

Prior to the examination, it must be verified that all necessary instruments are placed on the sterile work table and that any additional instruments that may be needed are available in the room. All instruments should be checked for adequate function before positioning of the patient.

The size of introducer sheath to use must first be decided. If a diagnostic cystoscopy is planned, smaller sheaths, 16–21 Fr, are adequate. If a biopsy is planned, a larger sheath, 21–24 Fr with a biopsy channel, must be introduced. The irrigation fluid is then attached to the introducer sheath. The fluid bag should be opened so that the flow can be regulated with the small switch at the cystoscope. The obturator is inserted after the tip of the instrument is lubricated to facilitate passage through the urethra. For most procedures, a water-soluble lubricant such as Surgilube® is sufficient. If the urethra has to be inspected or in cases of mild stenosis, the cystoscope can be inserted without the obturator under careful direct vision. Special obturators are available, which allow visualization during introduction and provide a smooth surface[2].

Cystourethroscopes come with a variety of lens systems. Selection should be made prior to the start of the examination. If possible, the light source should be attached before introduction of the scope, to reduce handling of the instrument and patient discomfort as much as possible during the examination. A 0° lens is necessary for evaluation of the urethra and the bladder neck. A 30° lens is the most versatile for inspection of the bladder.

Orientation

Orientation is usually not difficult unless there is significant distortion. The first landmarks encountered are the ureteral orifices and the interureteric ridge at the trigonum. Often, these structures are closer to the bladder entrance than expected. The opposite landmark is the gas bubble at the top of the bladder. Of course, one should always keep in mind the type of lens in use. It may be difficult to find the gas bubble with a 120° lens!

Course of the examination

Before the examination is initiated, the instruments should be ready. After inspection of the external genitalia and the urethral meatus, the scope should be inserted. As the female urethra is short, insertion is usually easy. The normal female urethra is similar to a French catheter, around size 18–22. The instrument should slide gently in without undue resistance. Force should not be used. In patients with an atrophic, narrow urethra, dilatation may become necessary. However, as this can be quite painful, general or regional anesthesia should be considered. Good lubrication and gentle, controlled pressure is the key to success without injury. The bladder is reached after a few centimeters. One usually tends to insert the scope too far, leaving a small area of bleeding on the posterior bladder wall. After the introducer sheath is inserted into the bladder, the obturator is removed. Urine now flows freely and can be collected for culture or cytology. The lens system is then inserted. It must be ensured that all connections are tightly closed.

If urethral evaluation is indicated, this should be accomplished first, as all manipulations cause erythema of the mucosa rendering later evaluation for inflammation impossible. For this, a sheath with a short oblique beak should be selected to reduce the spillage of distension medium and facilitate visualization of the entire urethra. The beak should be inserted anteriorly to elevate the anterior urethral wall and keep it from collapsing over the objective. A 0° scope is used most frequently, although a 30° scope is also adequate. With insertion, the distension medium is allowed to flow into the urethra at about 75 ml/min for optimal visualization. While the distal urethra opens easily to the infusion pressure, the mucosal folds of the mid-urethra will not separate as easily, owing to the higher pressure. Obstructing the proximal urethral lumen during withdrawal sometimes helps to improve visualization of the urethral mucosa. In patients with incontinence and low urethral pressures, the lumen opens easily even in this area, facilitating the evaluation.

Outflow obstruction often involves the mid-urethra. With the scope about 1 cm distal to the bladder neck, the function of the proximal urethra can be observed during coughing, straining and voiding. Under normal circumstances, the position barely changes during these maneuvers. In patients with stress urinary incontinence, one may see hypermobility of the vesicle neck or descent of the urethrovesical junction. One may have to elevate the eyepiece of the scope for continued observation of the proximal urethra in these patients. In patients with an unstable bladder, the involuntary detrusor contractions opening the bladder neck can be observed especially during filling.

The urethral walls are inspected to detect signs of inflammation, ulceration, tumors, pseudomembranes, polyps, fronds, or the presence of exudate from urethral diverticula or the periurethral glands. The best way to detect secretions from the suburethral glands is to massage the urethra and the lateral tissues gently while withdrawing the 0° scope. The duct openings are seen on the floor of the urethra in the proximal third. One has to be very careful performing this maneuver, as the area can be extremely sensitive in cases of inflammation. Diverticula are demonstrated in the same manner in about 4% of patients. The majority originate from the mid- to proximal posterior urethral wall. These openings can be very small. One should look for crypts or indentations as a clue. In difficult cases, a right-angled lens may improve visualization.

Dynamic urethroscopy[3] can be helpful for the diagnosis of genuine urinary stress incontinence. The relationship between urethra and bladder during stress is directly observed. The bladder is filled with at least 200 ml of fluid. The 0° scope is withdrawn about 1 cm distal to the bladder neck. The central third of the viewing area will then be occupied by the partially closed internal meatus at the urethral vesicle junction. The patient is then asked to cough or perform Valsalva maneuvers while the internal meatus is directly observed for opening and closing. An incompetent bladder neck may remain open or close slowly during these maneuvers, whereas a competent bladder neck will usually remain closed. If there is uncontrollable opening during the bladder filling, detrusor instability should be suspected. Because there is controversy regarding the accuracy of this evaluation, a complete urodynamic evaluation is the preferred method of diagnosing stress incontinence. A more detailed description is beyond the scope of this text and the urological or urogynecological literature should be consulted.

For bladder evaluation, one may insert the sheath with an obturator in place and then change to the lens system, or proceed with placement under direct vision with the lens in place. Most patients can tolerate direct placement of the sheath and lens if downward pressure with the blunt beak of the instrument is used. For evaluation of the bladder, a 30° or 70° scope is recommended. In some patients, a 0° lens may also be adequate, and in rare cases of a large cystocele a 120° lens may be necessary. Visualization of the bladder base can be quite difficult in this situation. Inserting a finger into the vagina to reduce the prolapse may be helpful. The scopes can be changed without discomfort to the patient as the sheath remains in place. The bladder is filled with 200–300 ml of fluid or until the patient notices mild discomfort or the urge to void. Patients with interstitial cystitis or other chronic changes may be able to tolerate only a small amount of bladder distension. The scope is slowly turned to observe the side walls and the anterior wall of the bladder until the air bubble in the dome of the bladder is seen. The same movement is then made towards the other side. It is important to proceed in a systematic fashion during each examination to ensure a complete evaluation. Changes of the mucosa including vascularity, signs of chronic irritation and bleeding, trabeculation, diverticula, tumors, or extrinsic compression are noted. The exact position and size of a fistula should be described. Special attention should also be paid to the trigonum area.

Lastly, the ureters are identified on both sides. They are slit-like in appearance and lie about 2.5 cm apart on the interureteric ridge close to the trigonum. If the bladder is filled with water, the expression of urine can be noted quite well. To confirm this, indigocarmine can be injected and, in well-hydrated patients, efflux of dye from the ureteral orifices is noted in approximately 5 min. The position and function of the ureters should be noted and duplications should be documented. Often, function can be observed without dye application, but whenever a problem is suspected, especially during intraoperative cystoscopy, dye injection is recommended to confirm adequate function. After the examination is completed, the lens is removed and the bladder drained through the sheath.

The course of the examination is similar if the flexible cystoscope is used[4]. As mentioned previously, this examination is often less uncomfortable due to the smaller size of the flexible instrument. Also, the flexible scope usually permits complete bladder visualization without excessive or painful movements. On the other hand, the picture is not consistently as clear and bladder drainage is more difficult. This can be a significant disadvantage, especially if there is debris or blood in the bladder. With the newest technology, biopsies and even some operative procedures are now possible through the flexible cystoscope. As with sigmoidoscopy, though, the slope of the learning curve is not steep. Therefore, more complicated procedures should remain the domain of the specialist as extensive experience is needed in these difficult situations[5].

Biopsy

Large biopsies are best obtained in the operative suite. As a relatively larger, more uncomfortable sheath is needed, the procedure is painful and there is a risk of bleeding. These situations are more easily handled by a specialist during adequate anesthesia. However, a small biopsy rarely causes significant problems and can be performed during the routine examination. A slightly larger sheath has to be inserted that includes a small instrument channel. The outside opening is covered by a rubber tip to prevent leakage. The forceps is inserted gently, as it can easily bend. It is best to place the forceps perpendicular to the lesion to be biopsied – no force is needed. A very small sample, only a few millimeters in diameter, is taken. After performing the biopsy, the area is carefully irrigated and checked for hemostasis. Bleeding necessitating cautery is rare, unless a large biopsy is taken or significant infection or tumor are present. In these cases, biopsy may have to be postponed until the necessary equipment is available.

Intraoperative cystoscopy

Cystoscopy is an important component of the surgical treatment of urinary incontinence. It can be used to check the elevation of the urethral vesicle junction while tying the bladder suspension sutures. After complicated pelvic surgery, it is often the best way to rule out misplaced sutures inside the bladder or bladder laceration. A meticulous, systematic evaluation is necessary so that small injuries are not overlooked, especially if they are hidden within a mucosal fold. The presence of blood within the bladder also requires careful evaluation using sterile water as the distension medium. Misplaced sutures protruding into the bladder lumen or deforming the bladder wall can be detected and then removed under direct visualization.

Intraoperative cystoscopy also allows ureteral evaluation. Bloody urine is a sign of possible injury, as is tenting of the bladder wall close to the ureters. Ureteral function is evaluated after injection of 5 ml of indigocarmine. In most cases, dye efflux is noted within 5 min. Depending on the intravascular volume of the patient, excretion can be delayed up to 20 min. A fluid bolus and Lasix® may speed up the process. Significantly delayed or absent excretion is most commonly a result of ureteral kinking and rarely due to complete obstruction. Again, the offending sutures must be identified and removed. The placement of ureteral stents is critical if an injury is detected, to ensure adequate healing. If it is difficult or impossible to pass the stent in a retrograde fashion through the cystoscope, a small cystotomy should be made to allow direct insertion of a pig-tail stent or no. 5 pediatric feeding tube. The tube is exteriorized with the urethral catheter and pulled after healing is complete. Cystoscopy is necessary for removal of the pig-tail stent. If a pig-tail stent is placed, it is important to use an adequate length, reaching the renal pelvis and forming a complete coil in the bladder. Size 26 or 28 should be adequate. Stents that are too long irritate the bladder more and cause more symptoms. Stents that are too short tend to retract into the ureter.

After a vaginal procedure or if the patient has been placed in the ski position, it is easy to perform transurethral cystoscopy during surgery. However, access in patients in the supine position is difficult. At the end of a long and complicated procedure, one often feels too tired for time-consuming patient repositioning. Here, the suprapubic approach is recommended[6]. The bladder is filled through the transurethral catheter. The bladder dome can then be dissected free from an extraperitoneal approach. A 1–2-cm purse-string suture is placed with 2–0 Chromic or Vicryl® in the bladder wall. The entire bladder wall is then elevated and the bladder is entered with Metzenbaum scissors or a no. 11 scalpel. The cystoscope with a 30° or 0° lens is then directly inserted into the bladder without sheath or bridge. The purse-string suture is pulled tight to prevent leakage of fluid. The trigonum area is easily found under the transurethral catheter bulb. The ureters are slightly lateral and above. Adequate distension is necessary so that small folds do not obstruct the view. With the abdomen still open, injuries can be identified. After the scope is removed, a suprapubic catheter can be inserted through the same incision if necessary, or the bladder is closed by pulling the suture tight. No further treatment or prolonged catheter drainage is necessary unless indicated by the original surgical procedure. These small bladder incisions heal very well.

Evaluation for fistula

Rarely, a vesicovaginal or ureterovaginal fistula develops after complicated gynecological surgery. This is often due to a smaller laceration overlooked during surgery. Sometimes the bladder wall can be weakened, owing to extensive dissection or impaired blood supply in the postoperative period and necroses. Fistula formation can also occur after radical surgery for gynecological malignancy, and is reported in up to 2% of the patients undergoing radical hysterectomy. Pelvic radiation therapy can also lead to fistula formation even many years after the treatment. Radiation induces tissue fibrosis with decreased blood supply. If the number of vessels is further reduced because of age-related changes, the tissue is no longer adequately perfused and breaks down, leading to a fistula. Obstetric lacerations are very rare in the USA, but are the most frequent cause for fistula formation in developing countries.

The presenting symptom of continuous urinary leakage is obvious if a large defect is present, but may be more difficult to evaluate if the lesion is very small. Sometimes it is difficult to distinguish between significant incontinence due to urethral or bladder dysfunction. Evaluation for a small vesicovaginal fistula can be quite challenging. Adequate distension of the bladder is necessary. Methylene blue can be instilled into the bladder through a catheter and then detected in the vagina. Sometimes it is helpful to insert a tampon into the vagina and ask the patient to ambulate. If evaluation of ureteral function is necessary, indigocarmine can be injected and the excretion observed. If stain is also seen in the vagina, a ureterovaginal fistula is present. Oral medication with Pyridium® is another option. In this case, one looks for orange discoloration on the vaginal tampon. In addition, the bladder can be filled with methylene blue to diagnose a vesicovaginal laceration or combined fistula.

A cystoscopy is recommended for complete diagnosis, no matter what the cause of the fistula formation. A small fistula may be difficult to find, as it can be hidden behind a mucosal fold. Again, it is important that the bladder is adequately distended. Ureteral function can be observed. In addition, one gathers information about the relationship of the fistula to the ureters and the bladder base. This knowledge is vital for planning an adequate fistula repair.

Problem areas

Bleeding is rare after routine cystoscopy, even if a small biopsy is taken. Cystoscopy for significant hematuria or radiation injury, however, should always be performed in the operating room. Anesthesia is needed because the examination is very painful. A larger sheath is inserted so that clots can be removed and vigorous irrigation performed. The examination is often prolonged, as visualization is more difficult in the presence of bleeding and often has to be interrupted for irrigation.

Pain is not a significant problem during short, simple cystoscopy. Evaluation for interstitial cystitis should be done under anesthesia, as significant pain can be expected and a biopsy may be necessary. If mild pain medication is not sufficient, it is best to abort the examination. A patient in pain cannot cooperate and this increases the risk of injury.

Difficult visualization is usually due to bleeding and rarely due to debris. Again, these examinations are best performed under anesthesia, as more significant pathology can be expected and the examination may be prolonged. Changing the distension medium to sterile water or glycine is helpful.

A narrow urethra can cause pain and problems with insertion of the scope. In postmenopausal women, therapy with intravaginal estrogen for several weeks may be helpful. Sometimes, however, urethral dilatation is necessary.

When to quit

When severe pain is induced because of urethral stenosis, stricture, inflammation, or bladder pathology, the examination should be aborted. Re-evaluation under anesthesia, as well as consultation with a urological specialist, should be considered. The same is true if problems arise in passing the scope.

Cystoscopy for significant hematuria should usually be performed in the operating room by a physician comfortable in using cautery if necessary. It is better to abort an examination and reschedule than to risk a complication.

Sometimes during intraoperative cystoscopy, bleeding can obscure vision so that complete evaluation is not possible. In this case, it is better to proceed with a cystotomy placed in the dome of the bladder for open evaluation. Thus, the bladder can be irrigated more efficiently and complete visualization can be achieved. It is also very easy to pass stents using this approach. Again, these incisions heal well after slightly extended bladder drainage. The fistula rate is extremely low after a well-closed bladder incision, but high after an undetected bladder laceration. In the case of very dense adhesions

Table 1 Cystoscopy report

Patient Name	:		
Date of Birth	:	Clinic Number:	
Date	:		
Indication	:		
Pert. History	:		
Symptoms	: No	Yes:	
Allergies	: No	Yes:	
Medical Problems	: No	Yes:	
Patient Instructions Received:		Consent Signed:	
Abdominal Examination: Normal		Other:	
Vital Signs	: T: _____ R: _____	P: _____	BP: _____
Medications	: No	Yes:	
External Inspection	: Normal	Other:	
Images	: No	Yes	marked with (P)
Biopsy	: No	Yes	marked with (B)
Findings	: Normal Urethra, Bladder, Ureters		
	Other: _____		
Dynamic Urethroscopy: No		Yes:	
Result	:		
Comments	:		
Complications	: No	Yes:	
Tolerance	: Good	Bad:	
Time Started	: _____	Time Completed: _____	
Vital Signs Post-examination : R: _____		P: _____	BP: _____
Diagnosis	: Normal	Other:	
Medications Given	: No	Yes:	
Instructions	: Routine	Other:	
Return to Clinic	: _____	Next Examination: _____	
Endoscopist	:		

between the cervix, vagina and bladder, the dissection can also be done with the bladder open, decreasing the risk of injury in the trigonum area.

Complications

Overall complications after cystoscopy are rare. Infection is the most frequent, as can be expected after any instrumentation. Routine guidelines are followed for treatment.

Injury to the urethra or bladder is a rare problem mainly due to significant inflammation, tumor, or diverticula. Surgical repair may become necessary. Therefore, it is critical to avoid excessive force during the entire examination. Likewise, the bladder should not be overdistended.

Post-examination instructions

There are few post-examination instructions. The patient should take the prophylactic antibiotics if prescribed and consume adequate fluid for the rest of the day. Slight discomfort can often be treated with Pyridium. Mild hematuria, especially after a biopsy has been taken, is normal. The patient should notify the physician in the case of severe bleeding with passage of clots or the inability to void. Severe pain, fever, or chills should also be reported promptly.

DOCUMENTATION

As with any examination, adequate documentation is important. A preprinted form is the best way to assure completeness with a minimal amount of work. An example is shown in Table 1.

TRAINING

The technique of cystoscopy in female patients is not difficult to learn. After approximately 20 procedures, adequate proficiency is attained. Cystoscopy in male patients is more difficult and should be performed by a specialist who has adequate training and performs the procedure frequently. The main problem is not the technique, but the evaluation of the findings, and the decision as to when and where to biopsy. The physician should have the opportunity to perform the examination on a regular basis to remain proficient.

CONCLUSIONS

(1) Simple cystoscopy can be performed routinely using local anesthesia. In cases of severe bleeding, interstitial cystitis, or if more extensive manipulation is anticipated, general anesthesia is recommended.

(2) Prophylactic antibiotics may reduce the risk of infection.

(3) Glycine is the most versatile distension medium; however, for short diagnostic procedures, sterile water or normal saline is adequate.

(4) Evaluation of the urethra should be performed first and without local anesthesia if possible. Evaluation of the bladder should follow a strict plan to ensure completeness.

(5) Intraoperative cystoscopy should be considered after all urogynecological procedures and should be liberally used if there is any suspicion of urinary tract injury.

REFERENCES

1. Richardson D. Cystourethroscopy in urogynecology. *Urogynecology* 1989;16:817–25
2. Cundiff GW, Bent AE. Cystoscopy for the urogynecologist. In Ostergard DR, Bent AE, eds. *Urogynecology and Urodynamics: Theory and Practice*. Baltimore: Williams and Wilkins, 1995: 173–85
3. Robertson JR. Dynamic urethroscopy. In Ostergard DR, Bent AE, eds. *Urogynecology and Urodynamics: Theory and Practice*. Baltimore: Williams and Wilkins, 1995:166–9
4. Pavone-Macaluso M, Lamartina M, Pavone C, Vella M. The flexible cystoscopy. *Int Urol Nephrol* 1992;24:239–42
5. Beaghler M, Grasso M, Loisides P. Inability to pass a urethral catheter: the bedside role of the flexible cystoscopy. *Urology* 1994;44:268–70
6. Timmons MC, Addison WA. Suprapubic telescopy: extraperitoneal intraoperative technique to demonstrate ureteral patency. *Obstet Gynecol* 1990;75:137–9

Section III Sigmoidoscopy and Cystoscopy Illustrated

Robert F. Werkmann, Roger R. Dmochowski, Brigitte E. Miller,
Ward A. Katsanis and Greg Portera

List of illustrations

Figure 1.1
Basic controlhead design of video sigmoidoscope

Figure 1.2
Schematic of the tip of the endoscope shaft

Figure 1.3
Example of a room set up for sigmoidoscopy

Figure 1.4
Cleaning area: sequential plan

Figure 1.5
Anatomy of the anus and rectum

Figure 1.6
Variations of sigmoid positions

Figure 1.7
Technique for holding the scope

Figure 1.8
Impacted scope

Figure 1.9
Pulling back leads to better visualization of the lumen

Figure 1.10
Left and right turns can be achieved by bending the endoscope tip and twisting the shaft

Figure 1.11
Clockwise torque helps to pass the sigmoid

Figure 1.12
Moving around a bend is achieved by a combination of bending the tip of the instrument, gentle advancement and pulling back if necessary

Figure 1.13
N-loop in the center is straightened by clockwise torque or transferred into an α loop by counterclockwise torque

Figure 1.14
The techniques of finding the lumen

Figure 1.15
Insertion of the scope

Figure 1.16
Entering the descending colon

Figure 1.17
Normal anal canal, anal papilla and dentate line

Figure 1.18
Rectal polyp, dentate line

Figure 1.19
Normal rectal veins

Figure 1.20
Thrombosed external hemorrhoid

Figure 1.21
Small internal hemorrhoid

Figure 1.22
Solitary rectal ulcer syndrome

Figure 1.23
Ulcerative proctitis

Figure 1.24
Aphthous ulcer of Crohn's disease

Figure 1.25
Rectal mucosal tags of Crohn's disease

Figure 1.26
Rectovaginal fistula due to Crohn's disease

Figure 1.27
Perianal fissure due to Crohn's disease

Figure 1.28
Hyperplastic polyp

Figure 1.29
Sessile polyps

Figure 1.30
Pedunculate rectal polyp

LIST OF ILLUSTRATIONS

Figure 1.31
Adenocarcinoma of the rectum

Figure 1.32
Radiation proctitis

Figure 1.33
Normal colon

Figure 1.34
Melanosis coli

Figure 1.35
Pseudomembranous colitis

Figure 1.36
Moderate chronic idiopathic ulcerative colitis

Figure 1.37
Crohn's disease of the colon

Figure 1.38
Inflammatory or pseudopolyps

Figure 1.39
Chronic idiopathic ulcerative colitis

Figure 1.40
Ischemic colitis with ulceration

Figure 1.41
Diverticulosis with small lesions

Figure 1.42
Thickened mucosal folds associated with diverticula

Figure 1.43
Diverticulosis and diverticulitis

Figure 1.44
Everted colonic diverticulum and polyp, pedunculate

Figure 1.45
Pedunculate polyp of the sigmoid

Figure 1.46
Arteriovenous malformation

Figure 1.47
Obstructing adenocarcinoma of the sigmoid

Figure 2.1
Basic cystoscopy set

Figure 2.2
Assembled cystoscope

Figure 2.3
Diagram showing the basic design of a flexible cystoscope

Figure 2.4
The handle of a flexible cystoscope

Figure 2.5
The flexible cystoscope in comparison with a rigid instrument

Figure 2.6
Small flexible biopsy forceps inserted into the cystoscope

Figure 2.7
Biopsy forceps fixed to the sheath

Figure 2.8
Anatomy of the bladder

Figure 2.9
Cystoscope ready for use

Figure 2.10
Insertion of the cystoscope

Figure 2.11
Suprapubic approach to intraoperative cystoscopy

Figure 2.12
Normal female urethral meatus

Figure 2.13
Mucosal prolapse

Figure 2.14
Squamous carcinoma of the urethra

Figure 2.15
Normal urethral mucosa

Figure 2.16
Inflammatory polyps

Figure 2.17
Urethral inflammation

Figure 2.18
Bladder neck

Figure 2.19
Urethral diverticulum

Figure 2.20
Trigone

Figure 2.21
Squamous metaplasia

Figure 2.22
Normal ureteral orifice

Figure 2.23
Abnormal ureteral orifice

Figure 2.24
Intraluminal ureteral lesions

Figure 2.25
Normal bladder mucosa

Figure 2.26
Mural vessels

Figure 2.27
Telangiectasias

Figure 2.28
Acutely inflamed bladder coexistent with chronic vesical inflammatory processes

Figure 2.29
Scarring of the urinary bladder wall

Figure 2.30
  Cystitis cystica

Figure 2.31
  Cystitis glandularis

Figure 2.32
  Early trabeculation

Figure 2.33
  Advanced trabeculation

Figure 2.34
  Diverticulum

Figure 2.35
  Air at bladder dome

Figure 2.36
  Distortion by sutures

Figure 2.37
  Vesicovaginal fistula

Figure 2.38
  Carcinoma *in situ*

Figure 2.39
  Early transitional carcinoma

Figure 2.40
  Transitional carcinoma

Figure 2.41
  Advanced transitional carcinoma

Figure 2.42
  Cervix cancer involving the bladder

SIGMOIDOSCOPY AND CYSTOSCOPY ILLUSTRATED

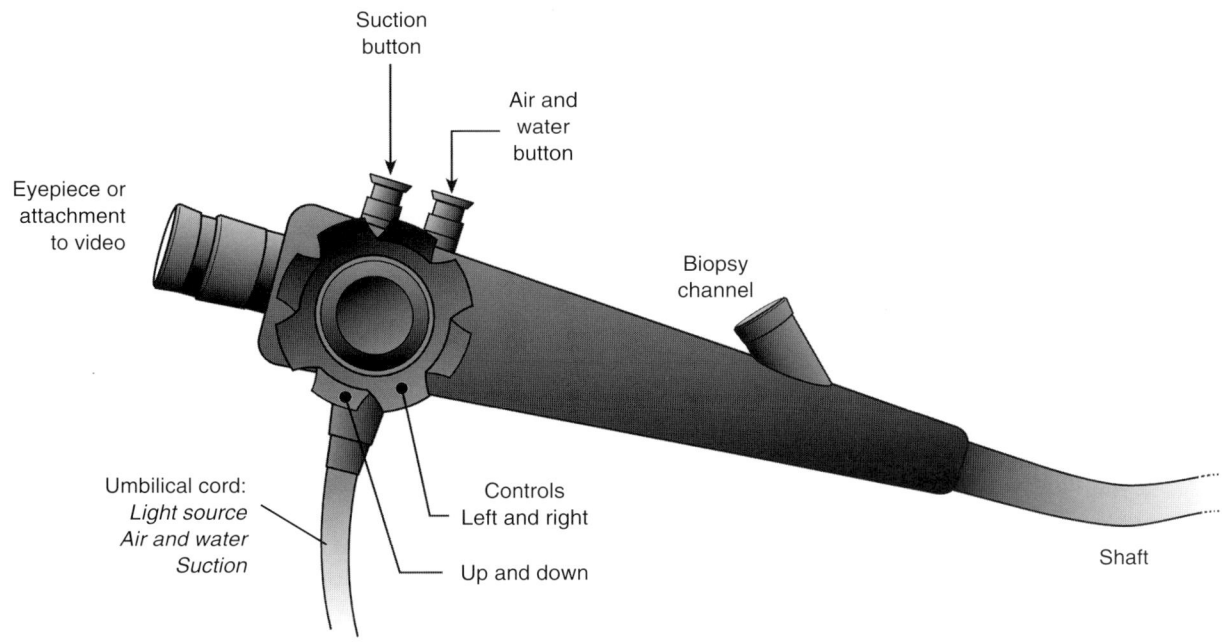

Figure 1.1 Basic controlhead design of video sigmoidoscope

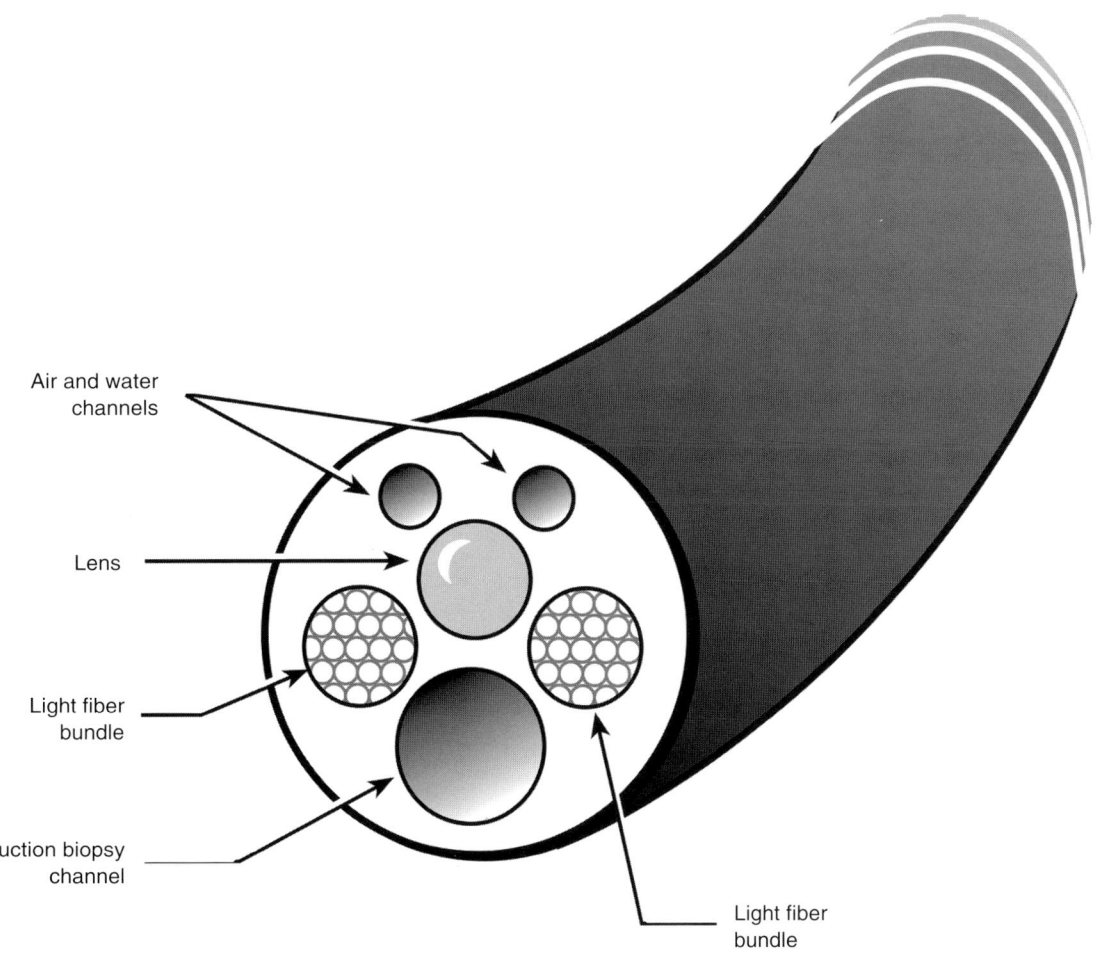

Figure 1.2 Schematic of the tip of the endoscope shaft

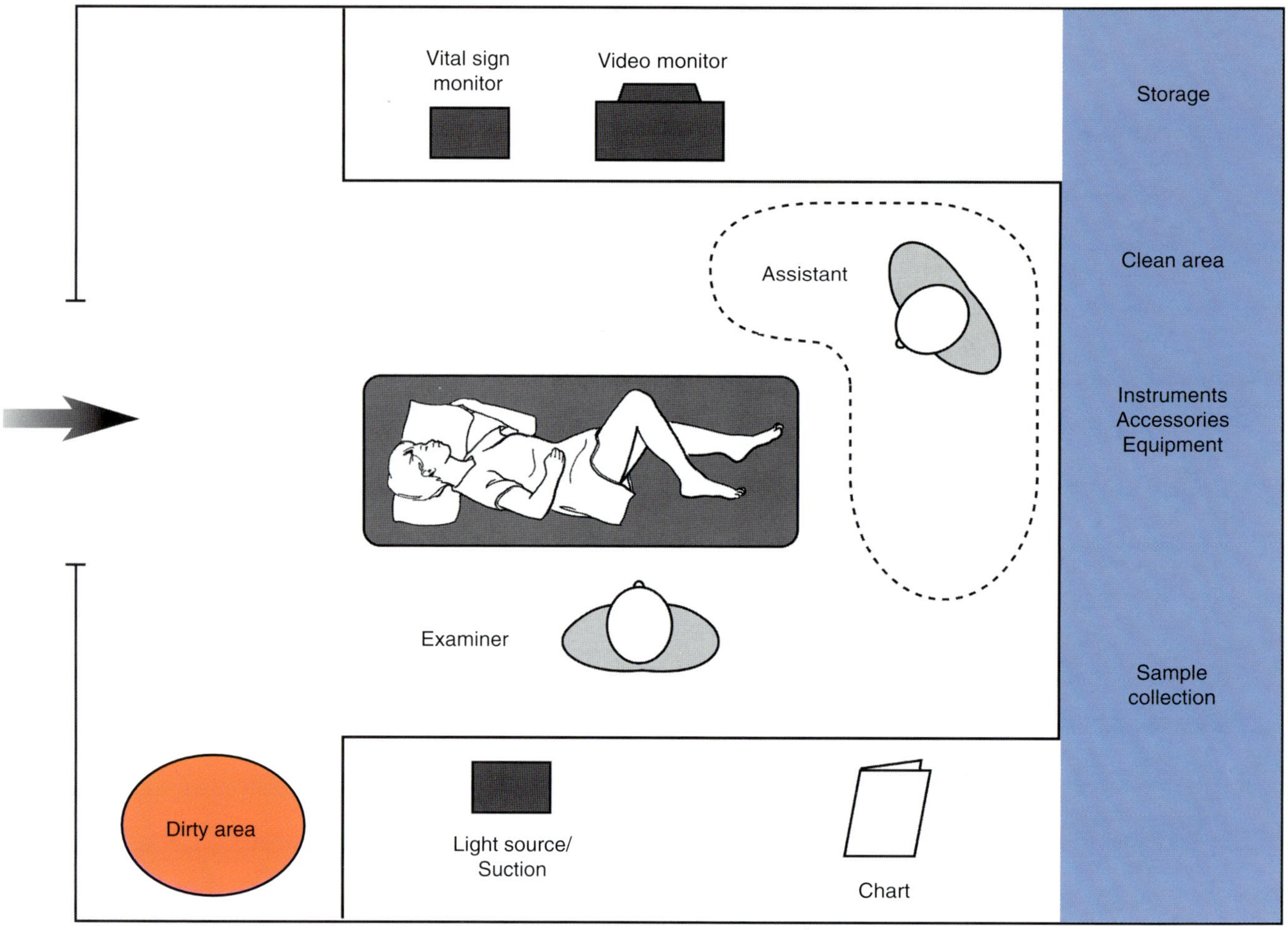

AN ATLAS OF SIGMOIDOSCOPY AND CYSTOSCOPY

Figure 1.3 Example of a room set up for sigmoidoscopy

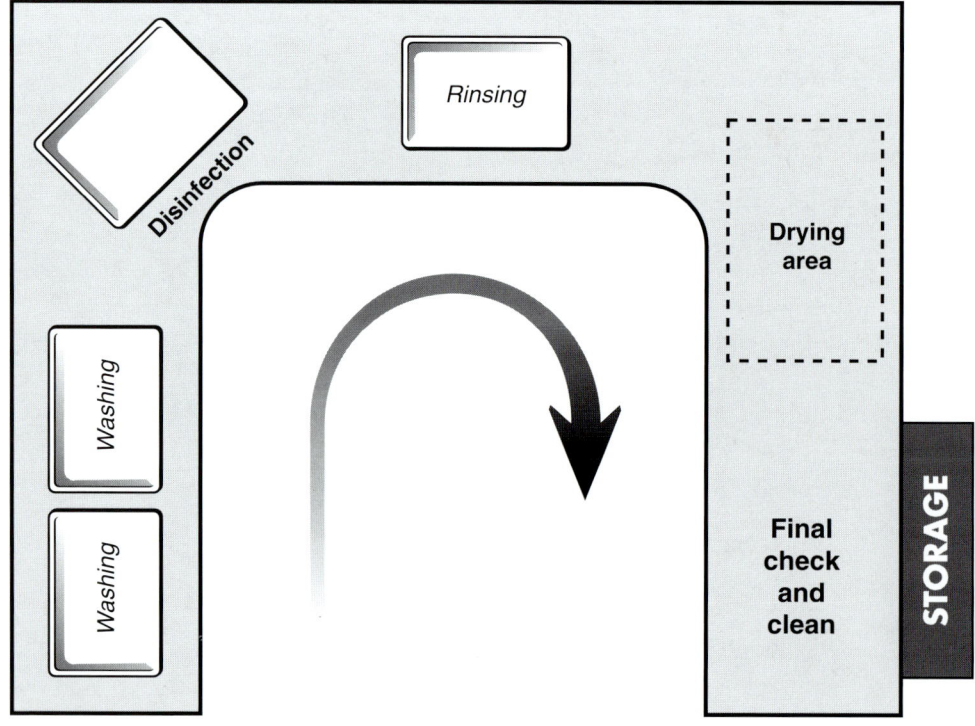

Figure 1.4 Cleaning area: 'sequential' plan

SIGMOIDOSCOPY AND CYSTOSCOPY ILLUSTRATED

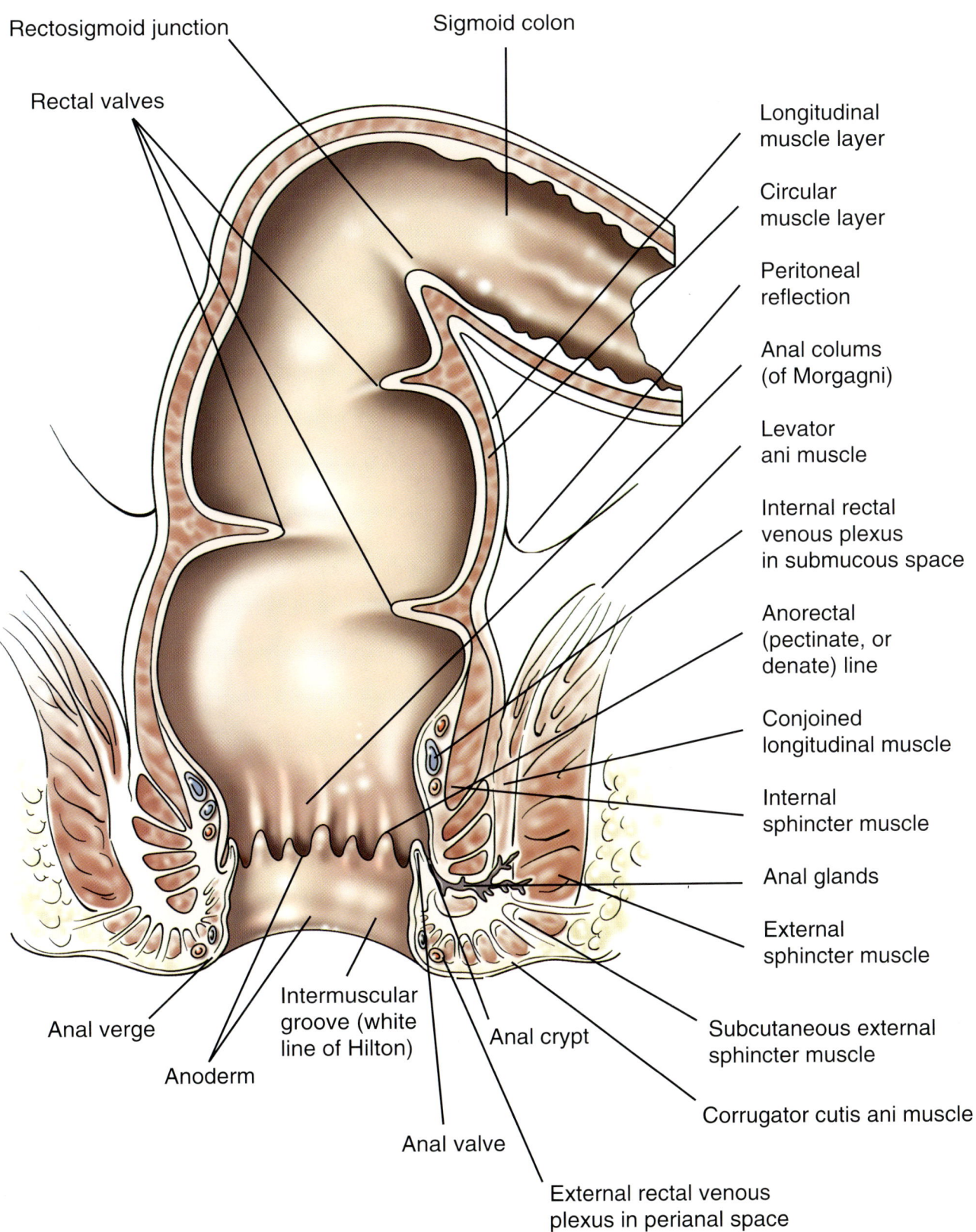

Figure 1.5 Anatomy of the anus and rectum

AN ATLAS OF SIGMOIDOSCOPY AND CYSTOSCOPY

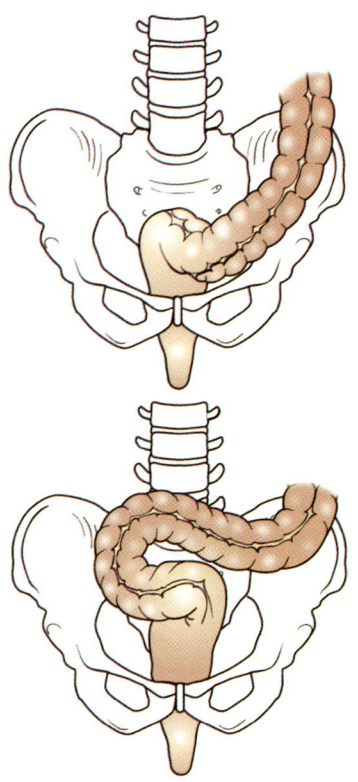
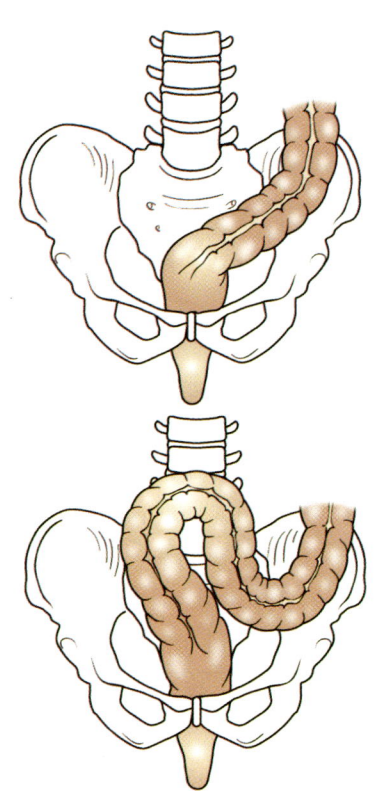

Figure 1.6 Variations of sigmoid positions

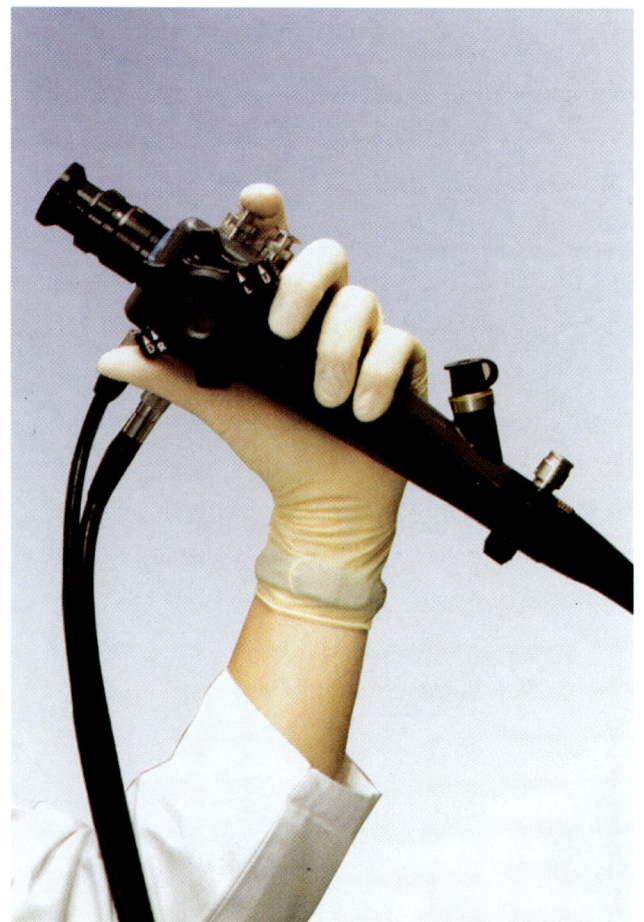

Figure 1.7 Technique for holding the scope. The instrument rests between the thumb and fingers. The first finger controls the air and suction. The thumb controls the up-and-down control wheel aided by the second finger

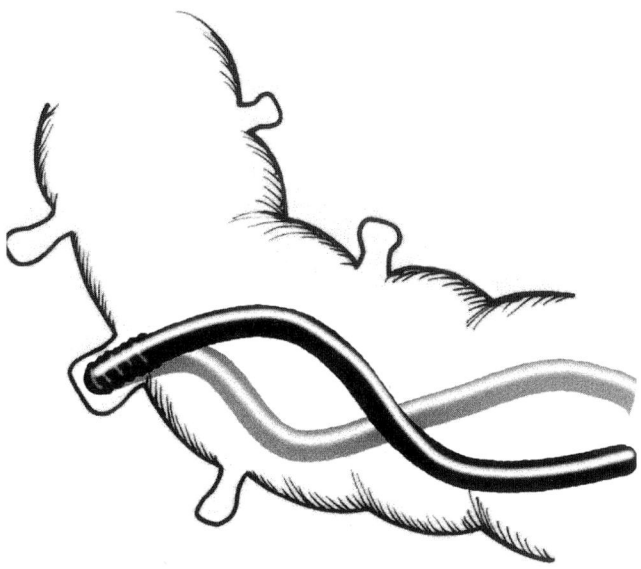

Figure 1.8 Impacted scope. Shaft movements cannot be transmitted to the tip. The scope has to be pulled back

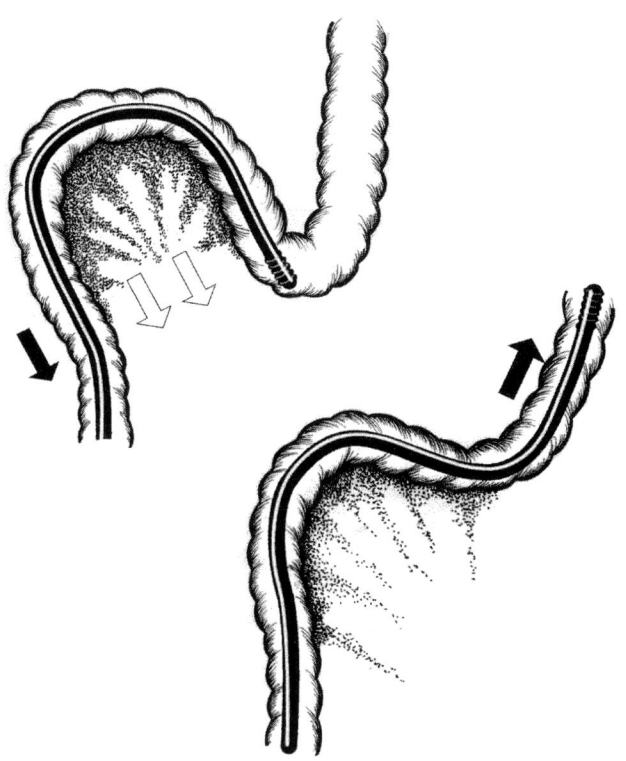

Figure 1.9 Pulling back leads to better visualization of the lumen and straightens the shaft, which renders it more responsive

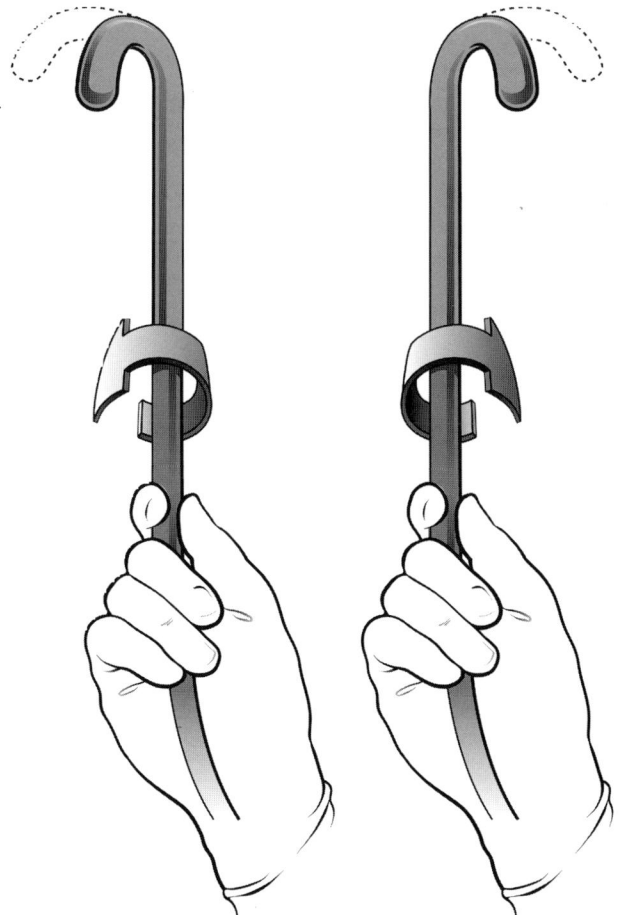

Figure 1.10 Left and right turns can be achieved by bending the tip and twisting the shaft to the right or left

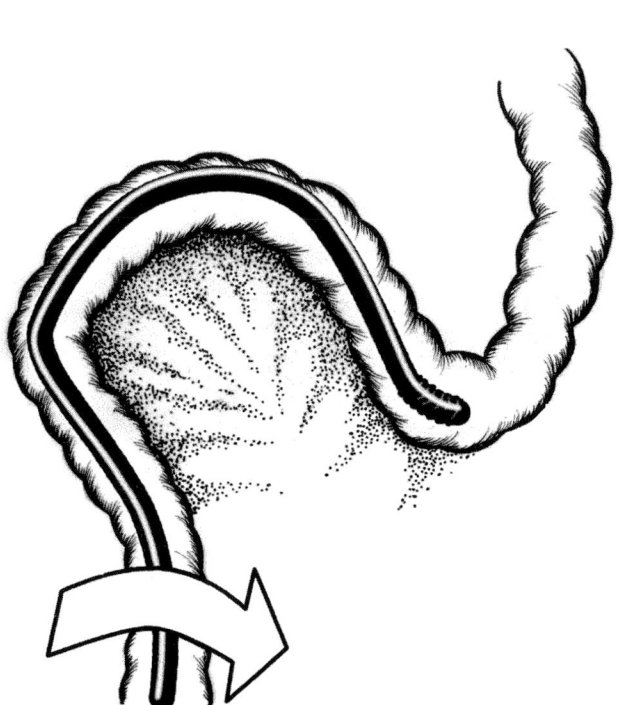

Figure 1.11 Clockwise torque helps to pass the sigmoid

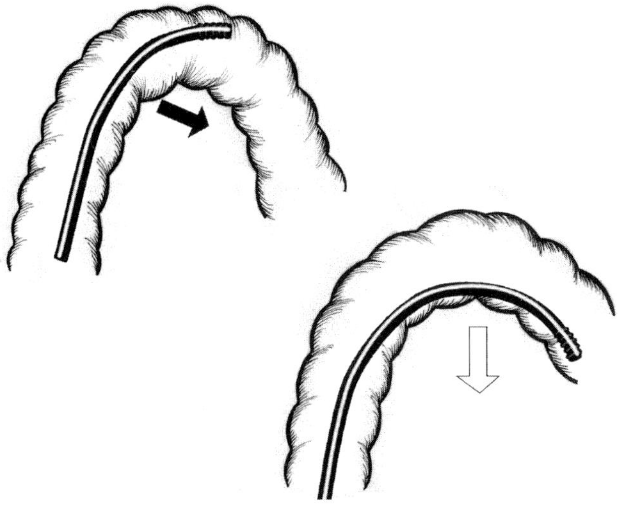

AN ATLAS OF SIGMOIDOSCOPY AND CYSTOSCOPY

Figure 1.12 Moving around a bend is achieved by a combination of bending the tip of the instrument, gentle advancement and pulling back if necessary

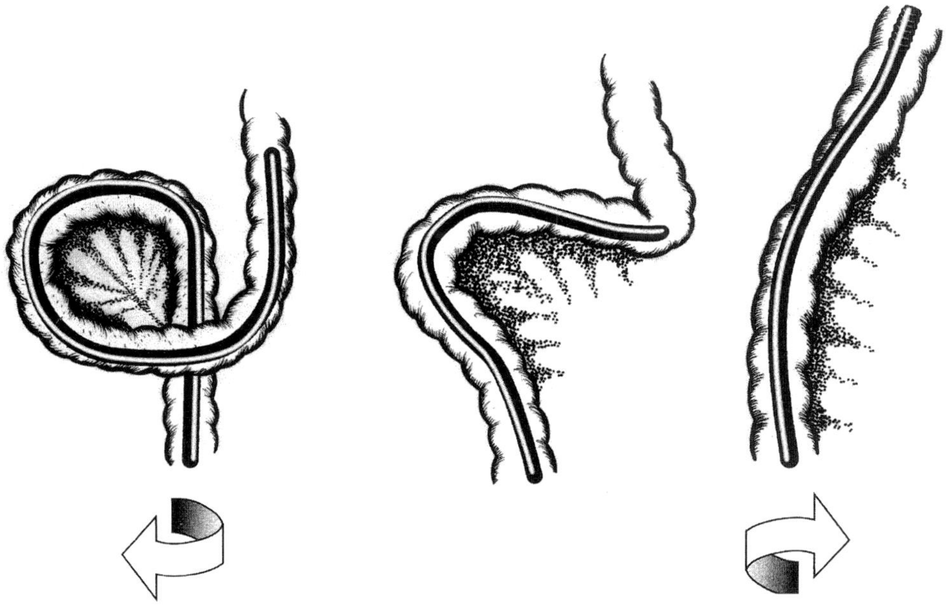

Figure 1.13 N-loop in the center is straightened out by clockwise torque or transferred into an α loop by counterclockwise torque to enter the descending colon

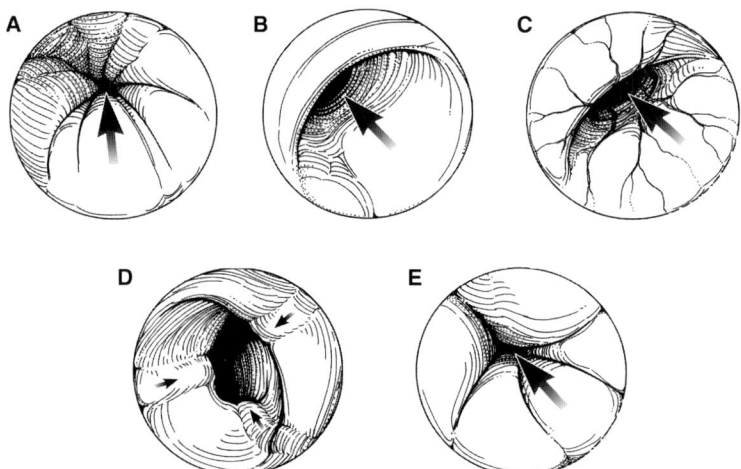

Figure 1.14 The techniques of finding the lumen. (A) Point the scope towards the darkest area to find the lumen. (B) The lumen is in the center of the arc formed by the haustral folds. (C) Mucosal vascular patterns point towards the lumen. (D) The taenia coli points towards the lumen. (E) If the bowel is deflated, the lumen is in the center of the converging fold

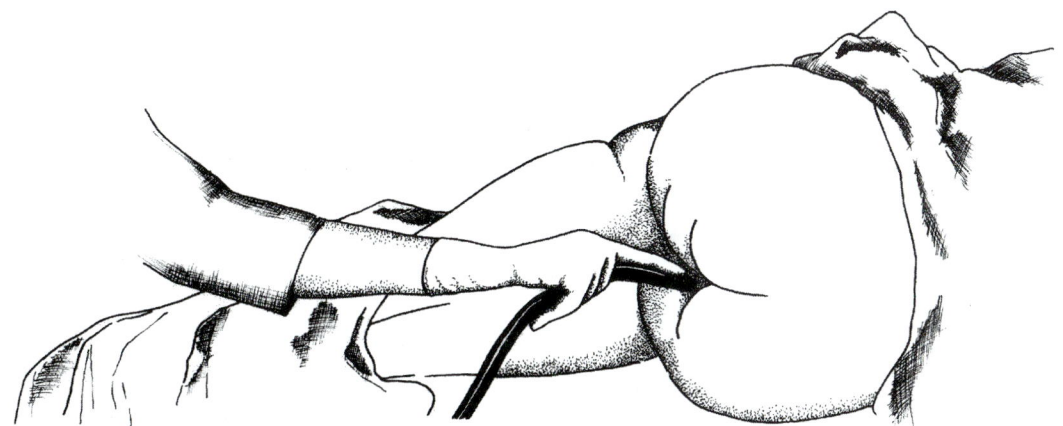

Figure 1.15 Insertion of the scope

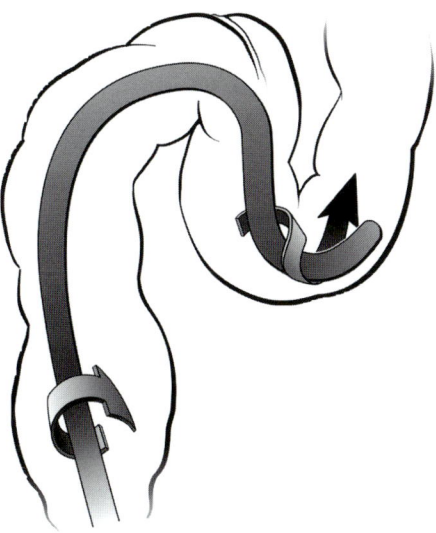

Figure 1.16 Entering the descending colon. Gentle pushing, little gas and mild clockwise torque are used to reduce the N-loop and keep the scope as straight as possible

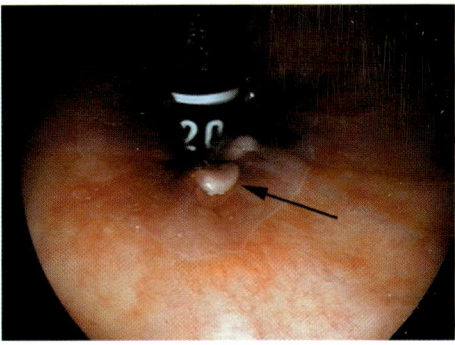

Figure 1.17 Normal anal canal, anal papilla, dentate line. These whitish raised nodules are present below the dentate line and exist as part of the normal skin of the external anal canal. There is normal sensory innervation of the dermis and hence accidental biopsy or endoscopic removal of this structure, seen here on retroflexed view, is painful and to be avoided

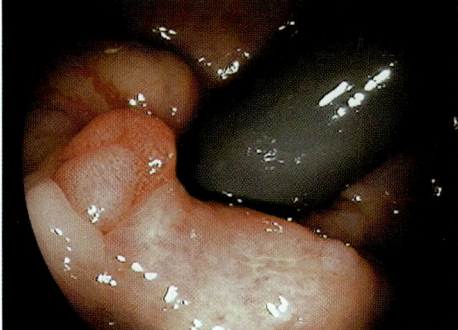

Figure 1.18 Rectal polyp, dentate line. Retroflexion of the endoscope is essential to prevent overlooking clinically significant neoplastic lesions of the rectum. This patient complained of intermittent rectal bleeding. This 0.5-cm adenomatous polyp was visible only on retroflexed view

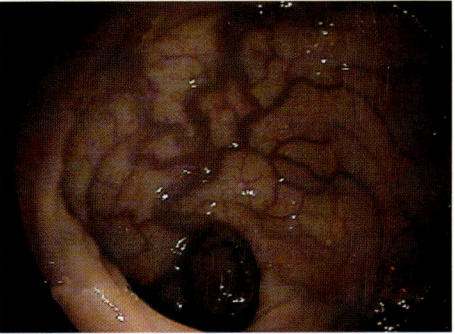

Figure 1.19 Normal rectal veins. These are prominent but normal rectal veins. In patients with portal hypertension, such veins may become rectal varices and produce life-threatening hemorrhage

AN ATLAS OF SIGMOIDOSCOPY AND CYSTOSCOPY

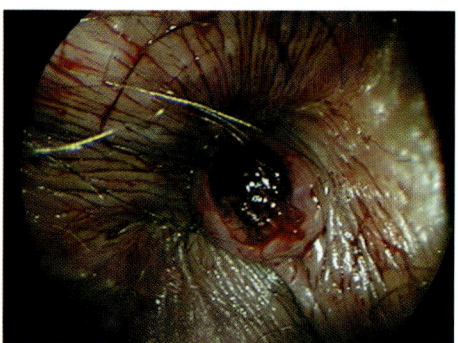

Figure 1.20 Thrombosed external hemorrhoid. This lesion truly is as painful as its appearance. A thrombosed external hemorrhoid has spontaneously become eroded, making excision of the clot with prompt pain relief straightforward. Careful visual inspection prior to performing anoscopy and flexible sigmoidoscopy will prevent an uncomfortable examination

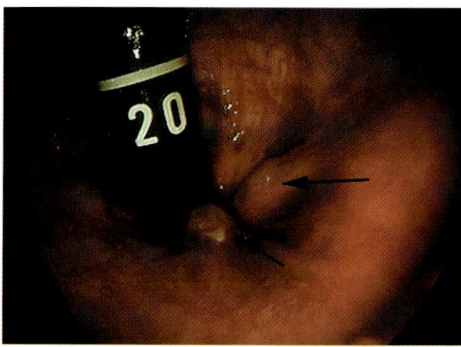

Figure 1.21 Small internal hemorrhoids. Small internal hemorrhoids noted on retroflexed view of the rectal vault

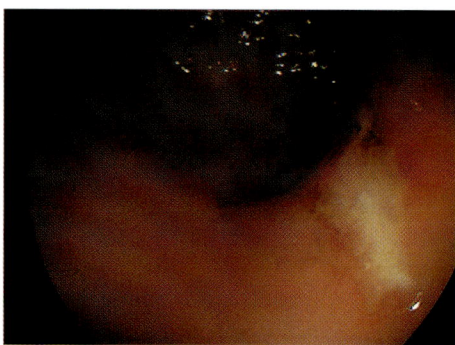

Figure 1.22 Solitary rectal ulcer syndrome. A posteriorly located superficial mucosal ulceration, often seen in patients with idiopathic chronic constipation, is evidence of the solitary rectal ulcer syndrome (SRUS). Chronic straining at stool from a functional pelvic outlet obstruction (levator ani syndrome) results in rectal mucosal prolapse and secondary mucosal ischemia, producing this chronic focal mucosal injury

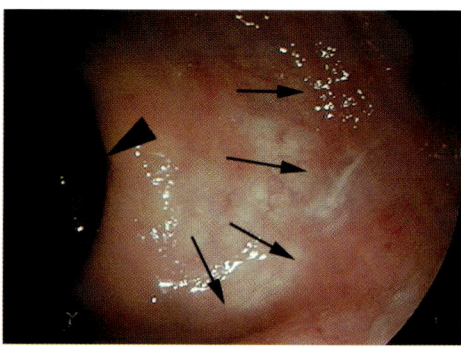

Figure 1.23 Ulcerative proctitis. Sharp demarcation between inflamed and normal rectal mucosa in a case of ulcerative proctitis. Note the presence of mucosal light reflexes on the left and their absence over the inflamed mucosa on the right. The proctitis is so mild that the submucosal vascular pattern remains partially visible

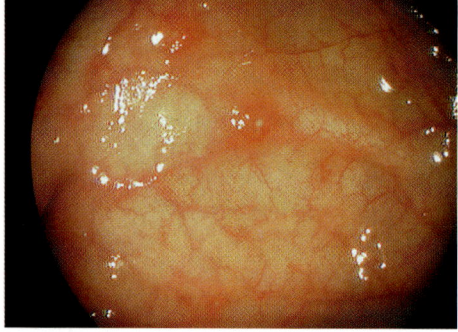

Figure 1.24 Aphthous ulcer of Crohn's disease. Punctate mucosal erosions, 1–3 mm, surrounded by a halo of erythematous edema, characteristic of the earliest mucosal lesion of Crohn's colitis, the aphthous ulcer. The surrounding colonic mucosa is normal

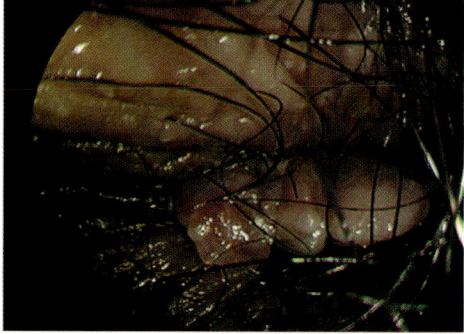

Figure 1.25 Rectal mucosal tags of Crohn's disease. Edematous protruding rectal mucosal tags, usually mistaken for prolapsed hemorrhoidal tissue, are one of the characteristic findings of perineal Crohn's disease. This pre-endoscopic physical finding demands an exhaustive search for evidence of gastrointestinal involvement with Crohn's disease

SIGMOIDOSCOPY AND CYSTOSCOPY ILLUSTRATED

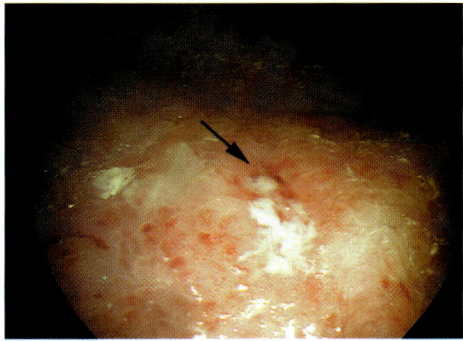

Figure 1.26 Rectovaginal fistula due to Crohn's disease. A rectovaginal fistula is a devastating complication of Crohn's disease in the female patient. The detection of the site of the fistula by endoscopic means is exceedingly rare. This flexible sigmoidoscopic examination immediately followed a double contrast barium enema, which demonstrated the communication with the posterior vaginal wall

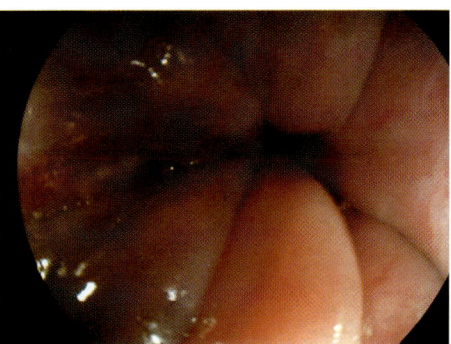

Figure 1.27 Perianal fissure due to Crohn's disease. Linear ulceration of the anal canal in a patient with ileocolonic Crohn's disease. Rectal bleeding and dyschezia prompted full colonoscopic evaluation

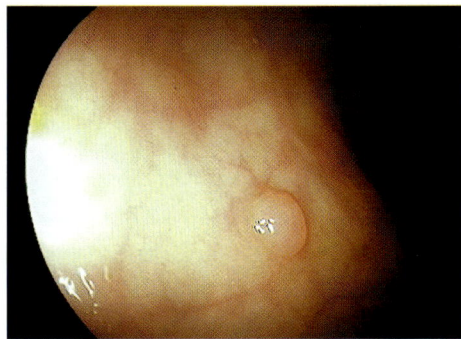

Figure 1.28 Hyperplastic polyp. Sessile rectal polyp, 2 mm in size. These small mucosal excrescences are commonly found in the rectum. Biopsy revealed a non-neoplastic polyp. Hyperplastic polyps are not associated with an increased risk of more proximal adenomas

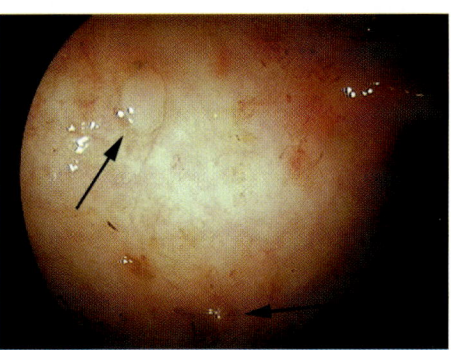

Figure 1.29 Sessile polyps. These sessile polyps were detected proximally to a large rectal cancer. Biopsy confirmed them to be tubular adenomas

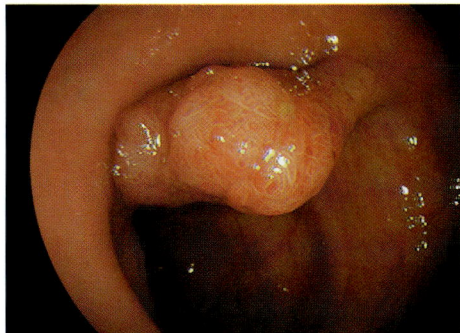

Figure 1.30 Pedunculate rectal polyp. A large, bilobed, pedunculate polyp in the rectum. This patient suffered from pruritus ani. Seepage of excessive mucus onto the perineum was due to the irritant effect of this large rectal lesion

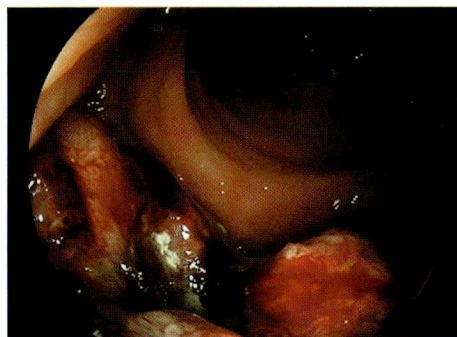

Figure 1.31 Adenocarcinoma of the rectum. Fungating polypoid mass, an adenocarcinoma, at the rectosigmoid junction. The rectal valves (of Houston) and rectosigmoid junction can be seen in the immediate background

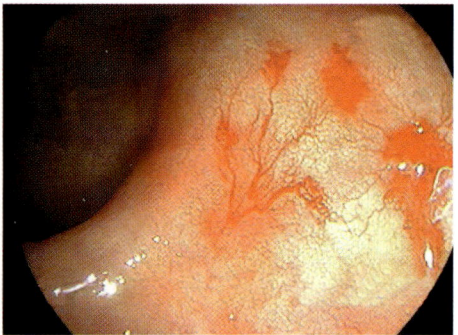

Figure 1.32 Radiation proctitis. Multiple telangiectatic vessels and arteriovenous malformations in the rectum of a patient who received radiation therapy for the treatment of cervical carcinoma. Radiation proctitis can be both acute and chronic. Acute symptoms include diarrhea with or without bleeding. Hematochezia is a common complaint in chronic radiation proctitis

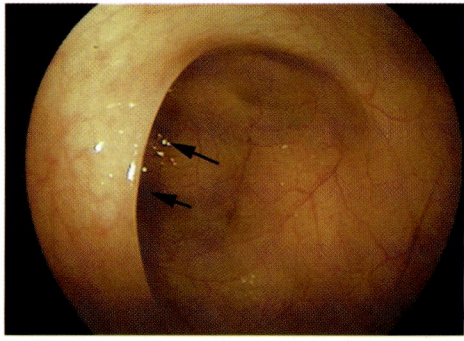

Figure 1.33 Normal colon. The normal colonic mucosa is pale and smooth. The lumen is found at the darker center of the arch

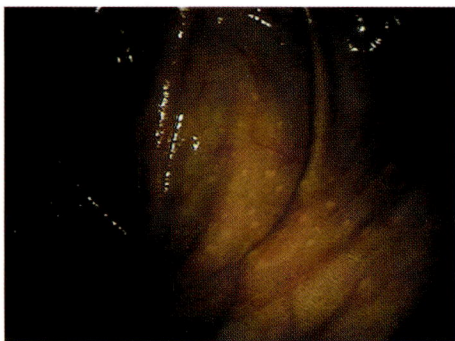

Figure 1.34 Melanosis coli. Erythema and granularity are seen when melanosis coli is part of the cathartic colon syndrome. Mucosal biopsy reveals melanin pigment granules in large submucosal macrophages, confirming laxative abuse. The dark brown reticulated pattern is seen throughout the colon in a patient using anthraquinone-containing laxatives

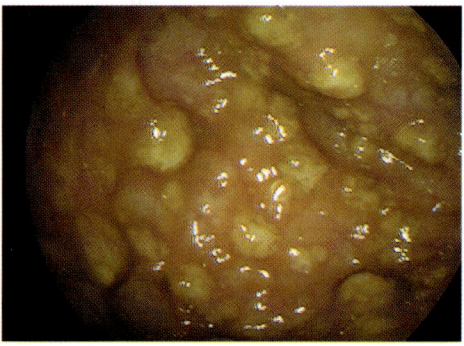

Figure 1.35 Pseudomembranous colitis. Grayish or bile-stained plaques of 1–5 mm (pseudomembranes). This discrete exudate of exfoliated mucosa, inflammatory cells and debris is classic for the severe form of antibiotic-associated colitis. The 'pseudomembranes' are typically left-sided and often spare the rectum, but the right colon only, or a pan-colonic distribution is not uncommon in severe disease. When the condition is detected in a patient with a diarrheal illness, the stool is usually strongly positive for *Clostridium difficile* toxin. With vigorous endoscopic water lavage, the gray-white appearance of adherent colonic pseudomembranes is distinctive

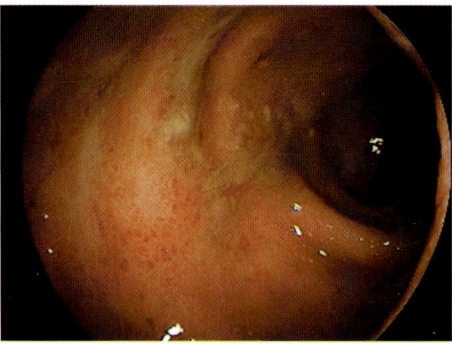

Figure 1.36 Moderate chronic idiopathic ulcerative colitis. This patient suffers from chronic idiopathic ulcerative colitis. There is loss of the mucosal light reflex and submucosal vascular pattern with edematous thickening of the haustral folds 'Mucopus' is visible in the colonic lumen

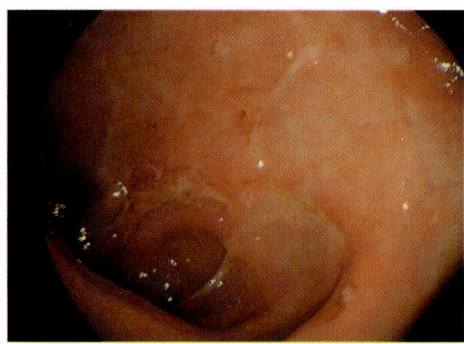

Figure 1.37 Crohn's disease of the colon. Colitis of Crohn's disease may be difficult to distinguish from chronic idiopathic ulcerative colitis (CIUC). Usually the rectum is spared

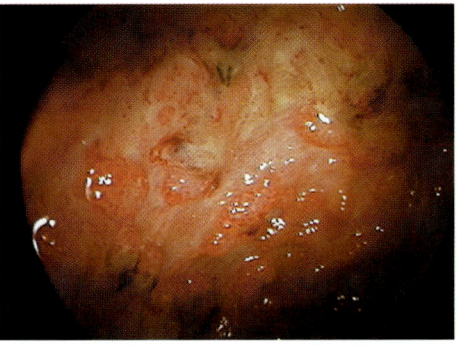

Figure 1.38 Inflammatory or pseudopolyps. Islands of regenerating colonic mucosa are frequently detected in both the quiescent and the acute resolving phase of chronic idiopathic ulcerative colitis (CIUC). Inflammatory polyps, so-called pseudopolyps, are non-neoplastic and do not have the characteristic histology of adenomas, but rather focal raised areas of granulation tissue with edema, fibrosis and neutrophilic mucosal infiltrate. They range in size from a few millimeters to several centimeters, are sessile and pedunculate, or may form mucosal bridges

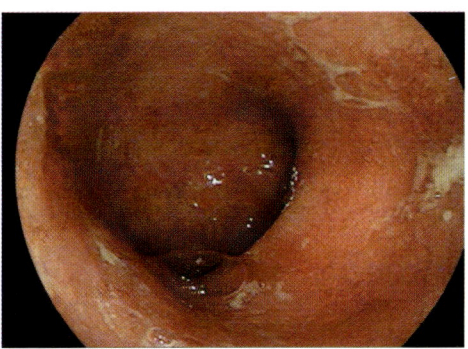

Figure 1.39 Chronic idiopathic ulcerative colitis. Loss of the light reflex and the submucosal vascular pattern due to the edema of mucosal inflammation suggest colitis. Diffuse microulceration and contact friability confirm colitis. Biopsy confirmed idiopathic ulcerative colitis

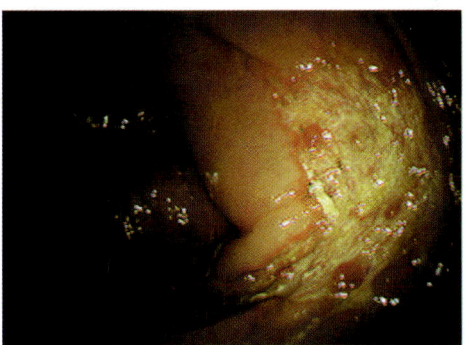

Figure 1.40 Ischemic colitis with ulceration. Stellate ulceration in a patient with segmental chronic ischemic colitis. Islands of granulation tissue are forming in the ulcer base

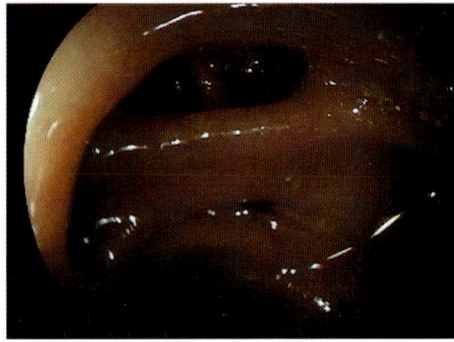

Figure 1.41 Diverticulosis with small lesions. Simple, multiple sigmoid diverticula

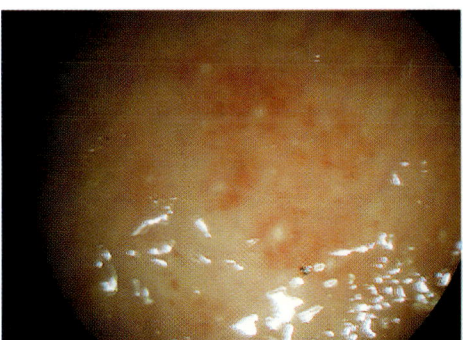

Figure 1.42 Thickened mucosal folds associated with diverticula. Thickened haustral fold pattern with patchy erythema, submucosal hemosiderin deposition and mild loss of the colonic vascular pattern is common in sigmoid diverticulosis. This appearance is frequently referred to as diverticulum-associated mucosal change. Chronic colonic muscular spasms with mucosal prolapse produce these subtle mucosal changes that may be mistaken for colitis. This endoscopic appearance may precede by years the development of the colon diverticula

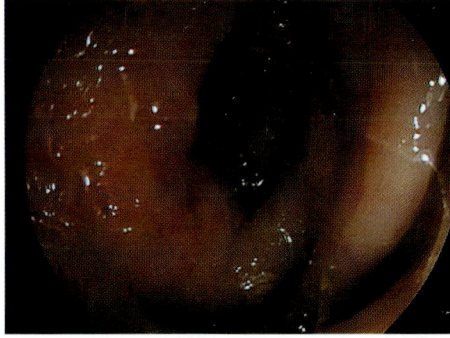

Figure 1.43 Diverticulosis and diverticulitis. Biopsy of a sigmoid colon mass in a patient suffering recurring bouts of diverticulitis. Exclusion of a precipitating colonic neoplasm is mandatory. In this case, the mass was simply acutely edematous and the inflamed mucosa was consistent with resolving diverticulitis

AN ATLAS OF SIGMOIDOSCOPY AND CYSTOSCOPY

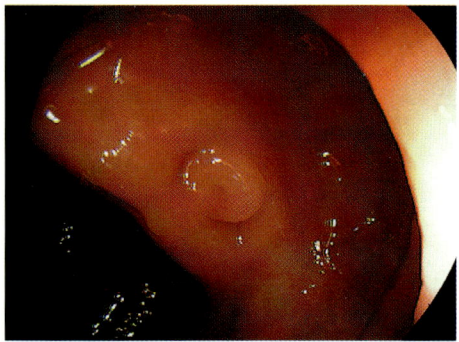

Figure 1.44 Everted colonic diverticulum and polyp, pedunculate. An unusual finding of an everted colonic diverticulum. It very much resembles a polyp. The tissue fold was soft and delicate when probed with the closed biopsy forceps

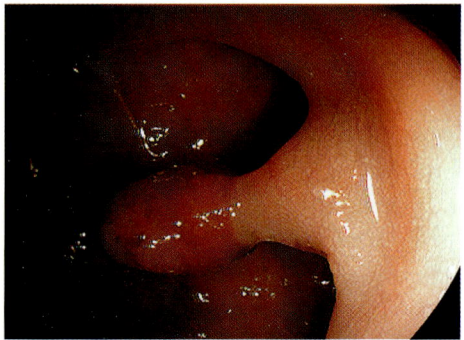

Figure 1.45 Pedunculate polyp of the sigmoid. A large, 15-mm, pedunculate polyp in the sigmoid colon. This patient should be referred for total colonoscopy with polypectomy. The polyp proved to be a tubulovillous adenoma

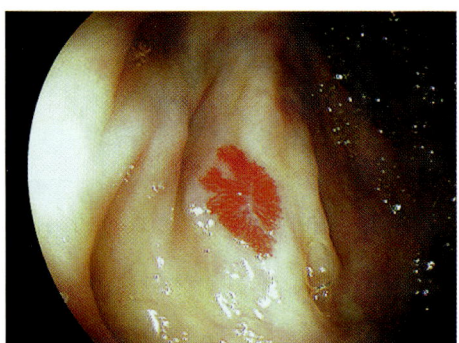

Figure 1.46 Arteriovenous malformation. A large, 10-mm, non-bleeding, colonic arteriovenous malformation. Usually asymptomatic and often multiple, they may occur anywhere in the colon. This is a frequent source of chronic lower gastrointestinal bleeding in the elderly

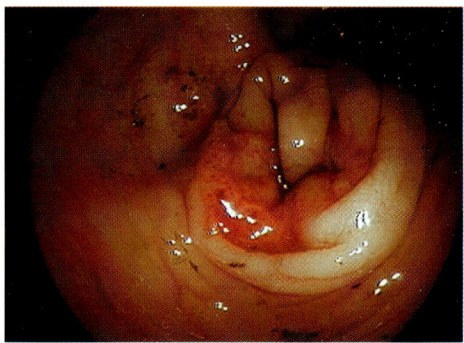

Figure 1.47 Obstructing adenocarcinoma of the sigmoid. Circumferential ('apple-core') obstructing sigmoid mass producing the common finding of a post-obstructive dilatation of the proximate distal colon. Biopsy confirmed moderately well-differentiated adenocarcinoma

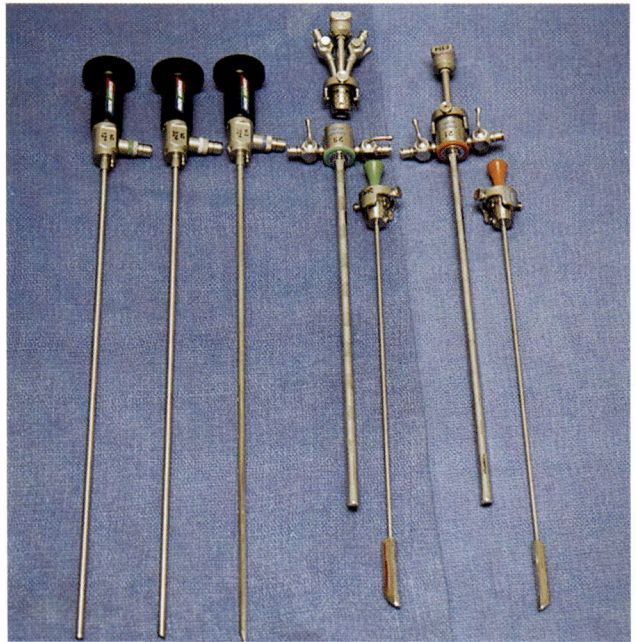

Figure 2.1 Basic cystoscopy set with telescopes 0°, 12° and 70°, sheath and bridge with biopsy channel, obturator, smaller sheath and bridge without biopsy channel and obturator

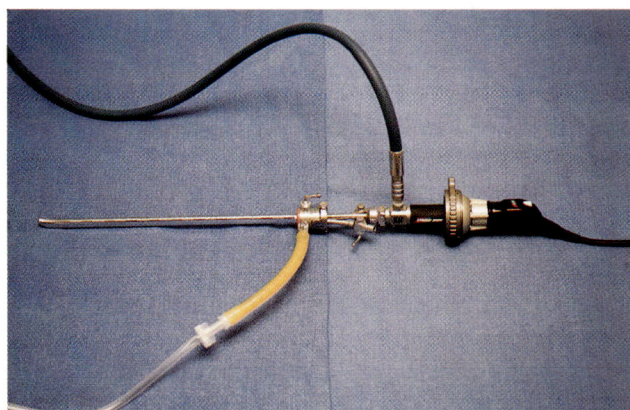

Figure 2.2 Assembled cystoscope with telescope in place, attachment for irrigation medium, closed biopsy channel, attachment for light source and camera

SIGMOIDOSCOPY AND CYSTOSCOPY ILLUSTRATED

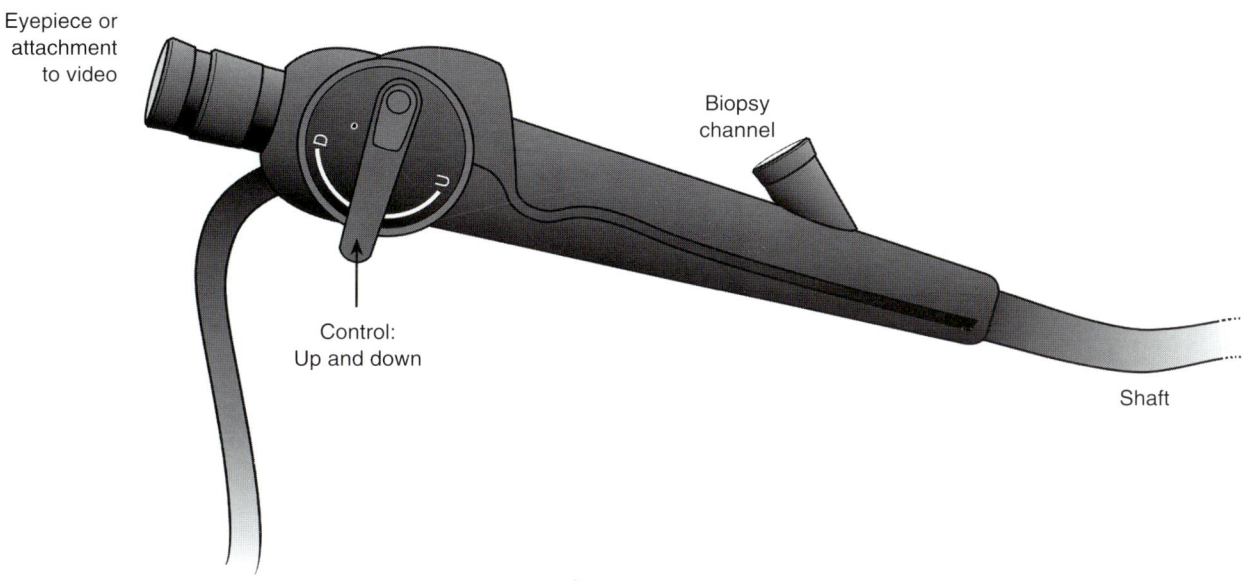

Figure 2.3 Diagram showing the basic design of a flexible cystoscope

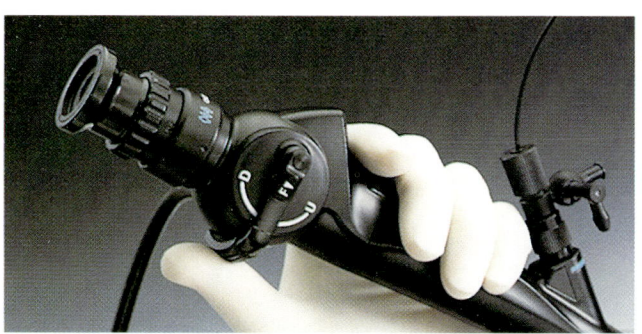

Figure 2.4 The handle of a flexible cystoscope showing the observation lens and the up-and-down tip control

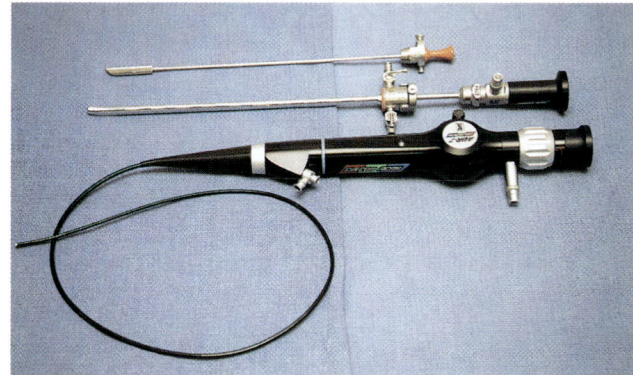

Figure 2.5 The flexible cystoscope in comparison with a rigid instrument

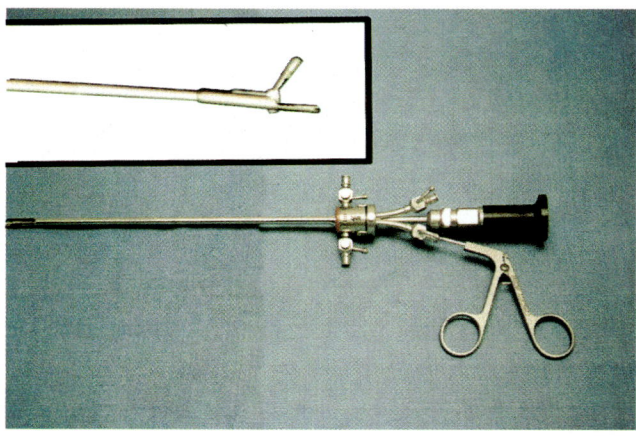

Figure 2.6 Small flexible biopsy forceps inserted into the cystoscope. The insert shows the tip of the forceps

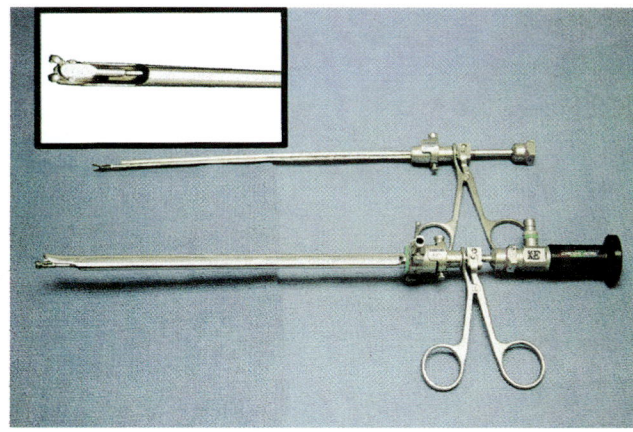

Figure 2.7 Biopsy forceps fixed to the sheath. Insert shows details of the forceps, which is larger, but is placed at a fixed angle

AN ATLAS OF SIGMOIDOSCOPY AND CYSTOSCOPY

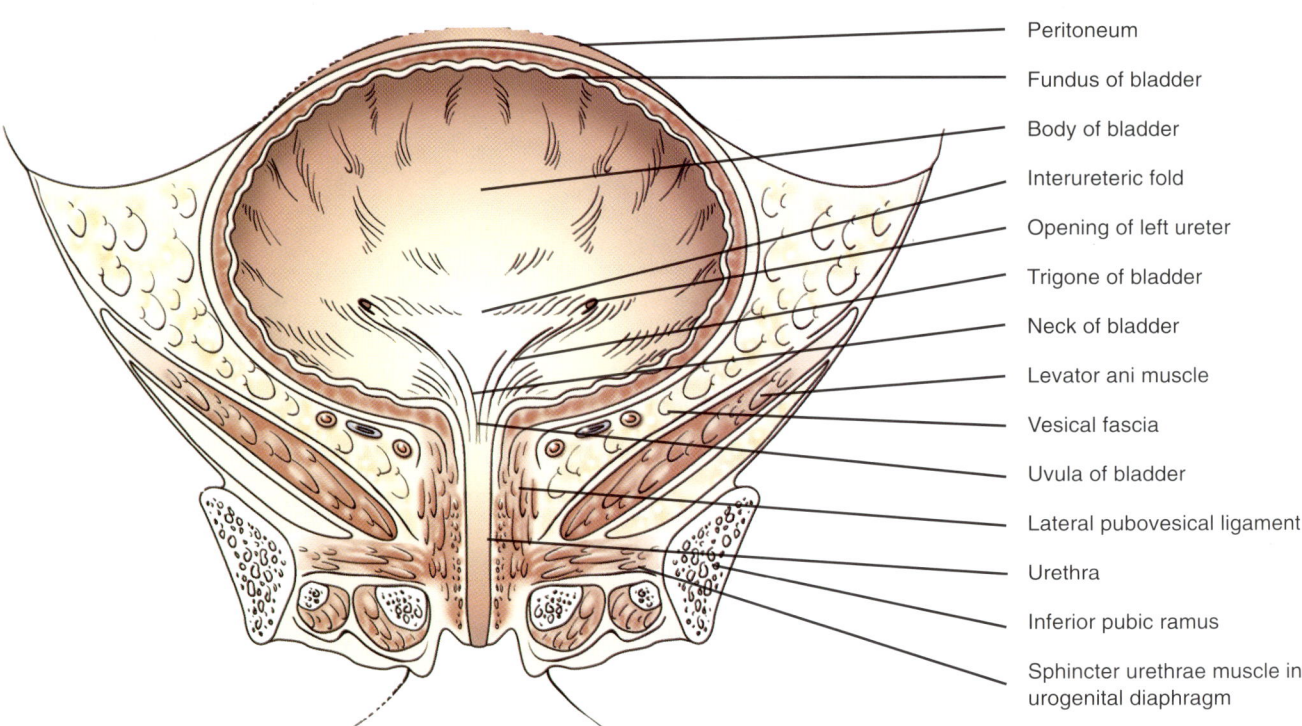

Figure 2.8 Anatomy of the bladder

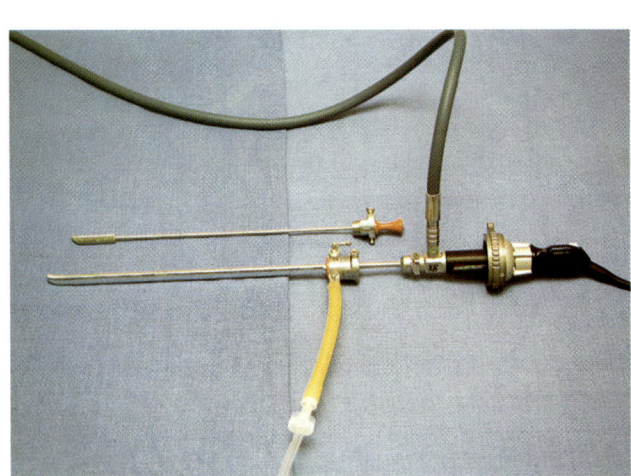

Figure 2.9 Cystoscope ready for use

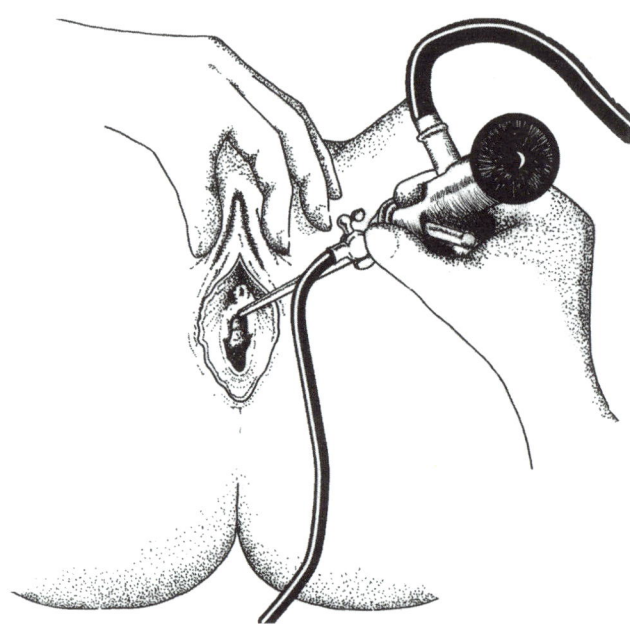

Figure 2.10 Insertion of the cystoscope. The labia minora are spread apart to decrease contamination

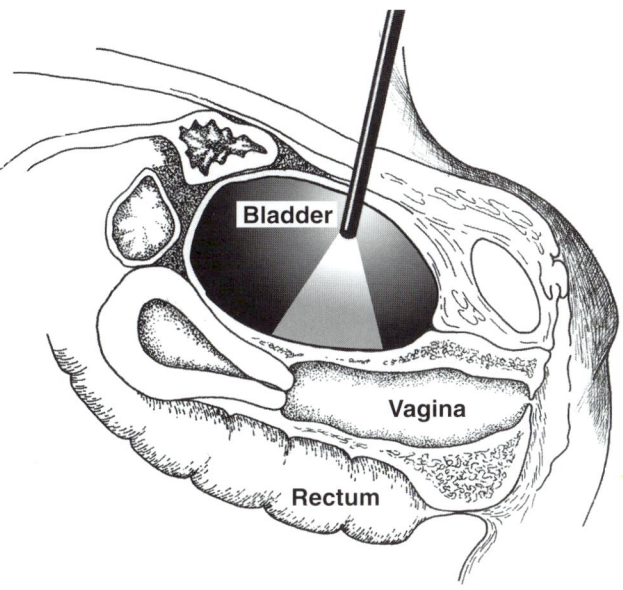

Figure 2.11 Suprapubic approach to intraoperative cystoscopy

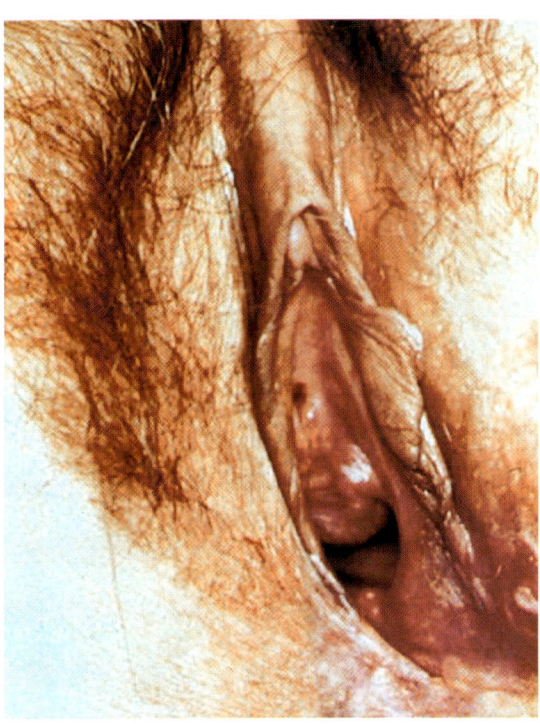

Figure 2.12 Normal female urethral meatus. The urethra is lined by transitional epithelium, which changes to the stratified vaginal epithelium near the orifice. The openings of the paraurethral ducts are at the lower border of the meatus. Atrophic changes of the vagina will also include the urethra

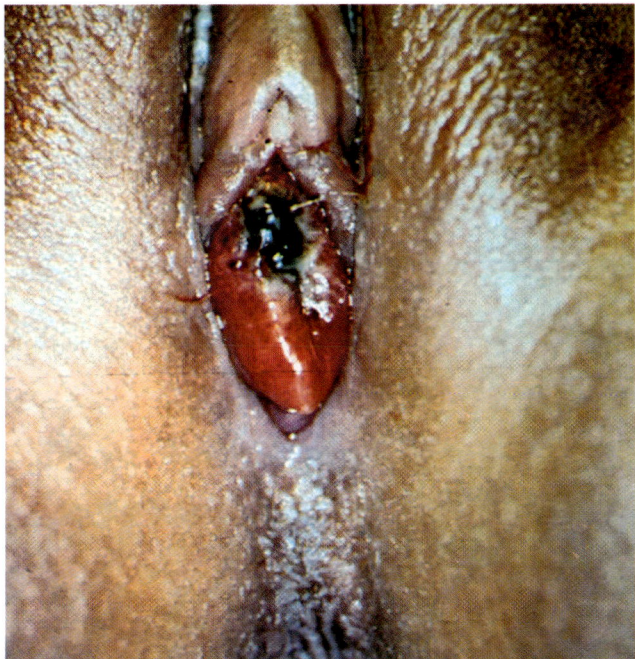

Figure 2.13 Mucosal prolapse. In the presence of significant atrophy, the urethral mucosa can become everted and then appears as a red tumor. Medication with estrogen cream is indicated in these situations

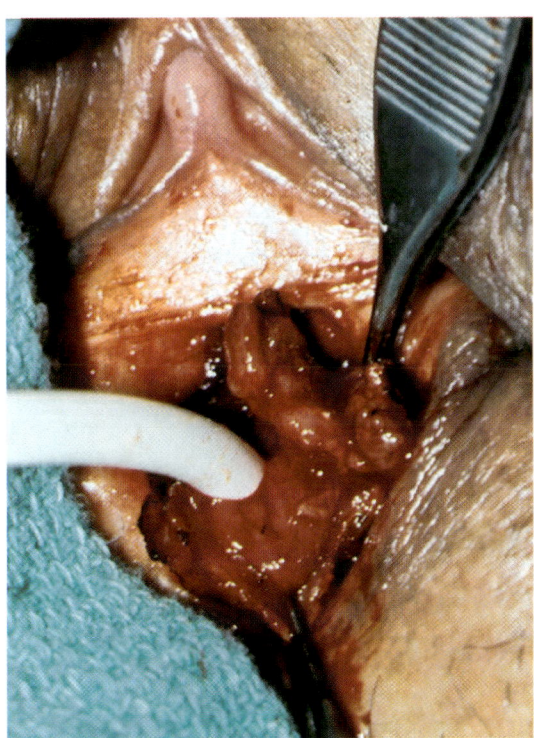

Figure 2.14 Squamous carcinoma of the urethra. Malignancies of the urethral meatus are rare. Most often, the meatus is involved with squamous cell carcinoma of the vulva. In contrast to mucosal prolapse, the tissue here grows irregularly and is very friable. In case of doubt, a biopsy is recommended; however, bleeding precautions have to be taken

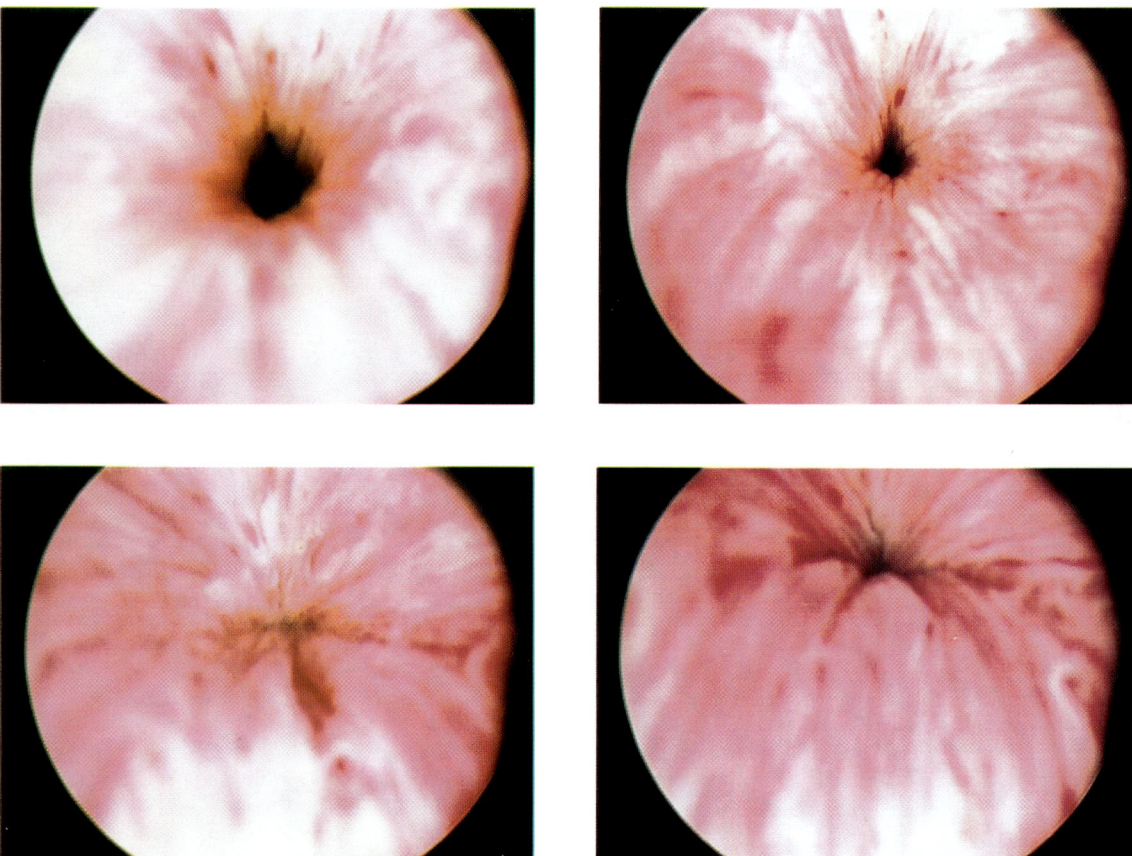

Figure 2.15 Normal urethral mucosa. The urethral mucosa and submucosa exert a 'seal' effect on the urethral lumen that is felt to be crucial to the continence mechanism. In menopausal and postmenopausal women this tissue is often atrophic and friable during instrumentation

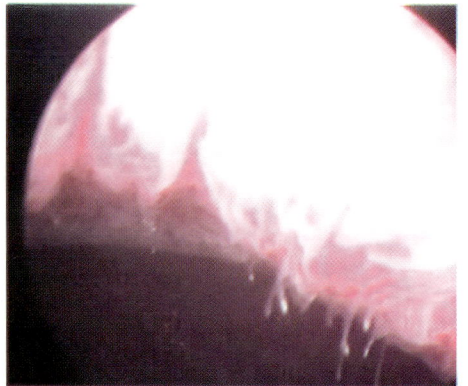

Figure 2.16 Inflammatory polyps. Urethral response to inflammatory agents such as infection, instrumentation, or atrophy can demonstrate a varied cystoscopic appearance. Inflammatory polyps are indicative of urethral mucosal hyperplasia that occurs in response to these inflammatory etiologies. Polyps may occur at any location within the urethra, but commonly are identified at the bladder neck, as in this case. These lesions are not malignant and, contrary to previous teaching, resection will not result in any symptomatic benefit to the patient

SIGMOIDOSCOPY AND CYSTOSCOPY ILLUSTRATED

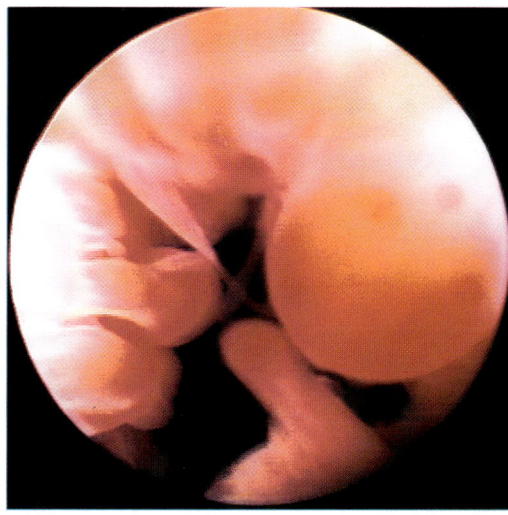

Figure 2.17 Urethral inflammation. Inflammatory changes may produce a quite dramatic appearance. Occasionally, the lesions may be confused with superficial urothelial malignancy or even create a mass effect. Polyps may completely involve the lining of the proximal urethra and bladder neck, as demonstrated in this photograph

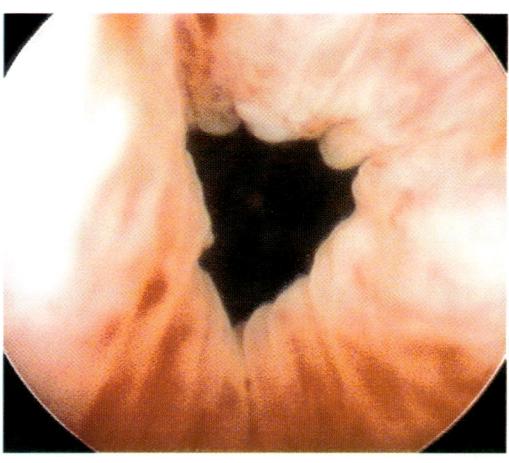

Figure 2.18 Bladder neck. The primary female continence mechanism resides at the bladder neck and is predominantly a smooth muscular mechanism that has extensive sympathetic adrenergic innervation. The bladder neck is easily identified as a prominent muscular ring at the level of the proximal urethra and bladder lumen

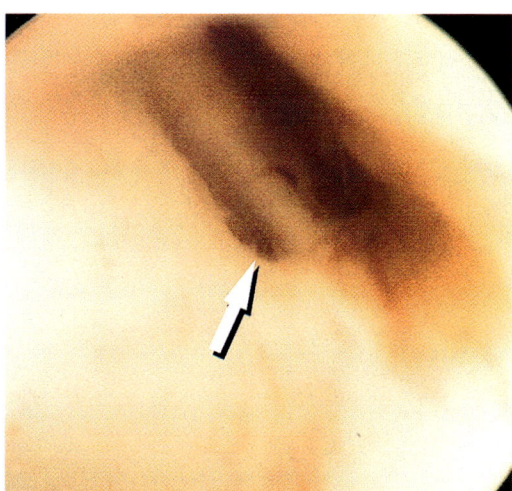

Figure 2.19 Urethral diverticulum. Cystoscopy can accurately identify urethral abnormalities. Various lesions are identifiable, either within the urethral lumen (stones, tumors, foreign bodies), or in the urethral wall (strictures, extrinsic compression). Careful examination of the urethral wall may identify the ostium of a urethral diverticulum (arrow). These apertures are most commonly noted on the floor of the urethra close to the bladder neck, often closely juxtaposed to folds in the urethral mucosa

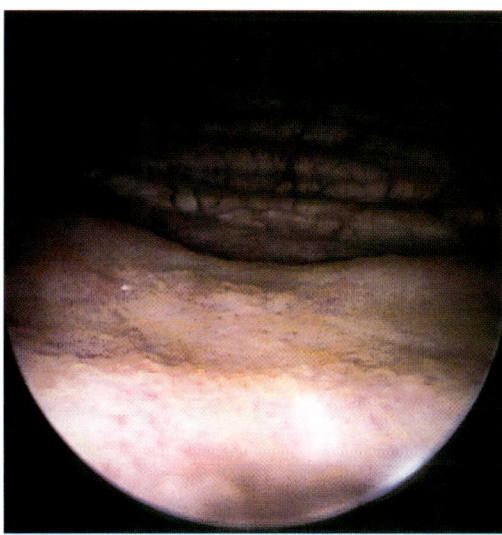

Figure 2.20 Trigone. The base of the bladder and urethra are profoundly sensitive to hormonal influences in women. The trigone (a triangular section of the bladder floor formed by the ureteral orifices laterally and the bladder neck medially) will demonstrate variable degrees of visual urothelial alterations that are dependent on the stage of the menstrual cycle and/or the menopausal status of the woman. In this example, a subtle irregularity of the bladder mucosa is demonstrated in the foreground. Behind this is a ridge that represents a muscular structure known as the intraureteric ridge (Bell's muscle)

AN ATLAS OF SIGMOIDOSCOPY AND CYSTOSCOPY

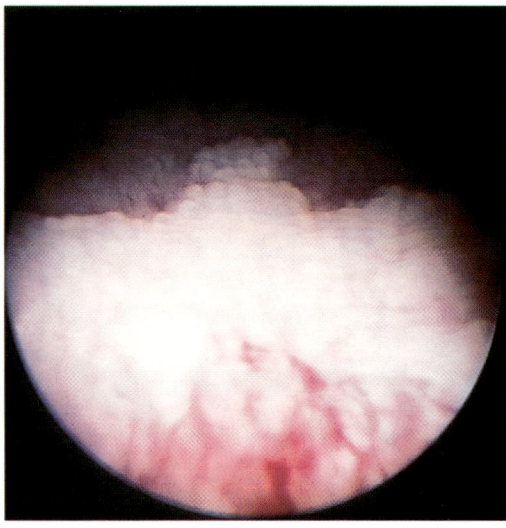

Figure 2.21 Squamous metaplasia. The trigone changes noted in Figure 2.20 can be quite prominent. The predominant histology in these areas of change is squamous metaplasia. The metaplastic change is whitish in color and has a cobblestone texture. Squamous metaplasia in this area can develop hyperplastic changes and can resemble an exophytic neoplasm. However, these hyperplastic squamous lesions usually occur within areas of flat squamous metaplasia

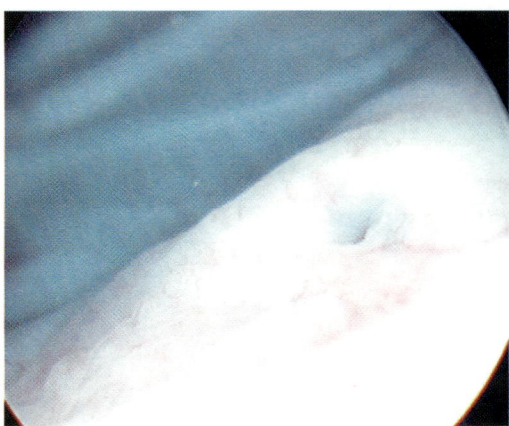

Figure 2.22 Normal ureteral orifice. A normal ureteral orifice has a flat or slightly raised contour and often will actively peristalse during observation. The location of the orifice is important to note. An orifice located on the lateral boundary of the trigone in parallel position to the contralateral orifice is described as orthotopic. An orifice occurring in any other location is considered ectopic and may be associated with either ureteral reflux or (less likely) obstruction

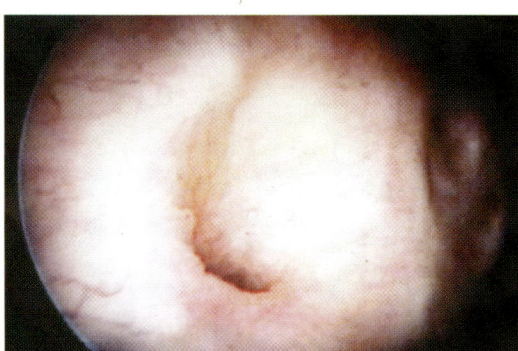

Figure 2.23 Abnormal ureteral orifice. An abnormally configured ureteral orifice may be indicative of natural pathology or iatrogenic injury. Primary bladder disease, such as neurogenic dysfunction (as in this case), congenital ureteral reflux or obstruction, or surgical fibrosis can distort an orifice. This anatomic distortion can also result in obstruction or reflux

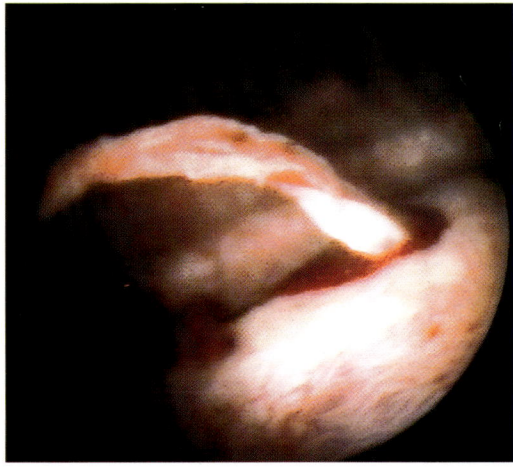

Figure 2.24 Intraluminal ureteral lesions. Intraluminal ureteral lesions can project from the ureteral orifice. Ureteral stones or tumors can cause distortion and displacement of the orifice and can also project into the bladder. Renal bleeding also may result in the transit of blood clots along the course of the ureter. Occasionally, a blood clot will lodge in the intramural portion of the ureter (that portion of the ureter that traverses the width of the bladder wall) and will project from the ureteral orifice. An organized clot can have a largely pale appearance, indicating an organized fibrin clot, as demonstrated

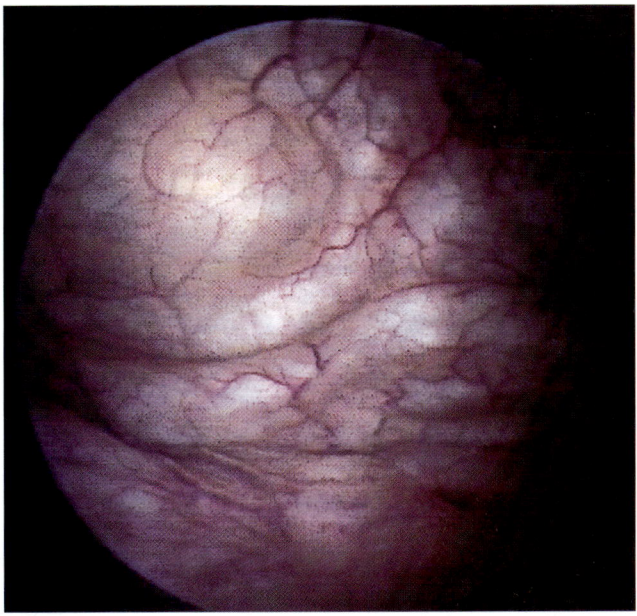

Figure 2.25 Normal bladder mucosa. Inspection of the bladder epithelium (or mucosa) must include visualization of the entire urothelial surface. The bladder mucosa should have a pink hue with prominent folds or redundancy of the lining seen. The submucosal vascular plexus of the bladder is easily discerned throughout the bladder. The vessels are delicate and resemble the branches of a tree. No other mucosal lesions should be identified

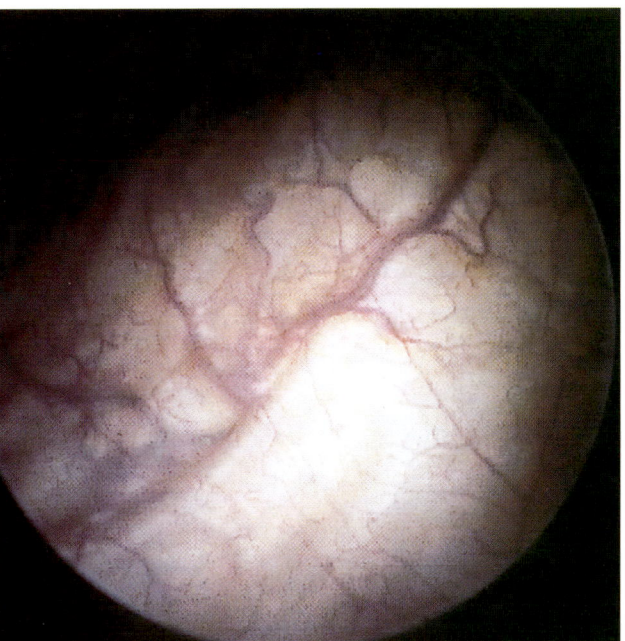

Figure 2.26 Mural vessels. Prominent mural vessels may be pulsatile and will often demonstrate numerable branches and a serpiginous course. Note that the vessel does not have any mural dilatation or abnormalities. In an acutely inflamed bladder (recent or ongoing urinary tract infection), the vessel may appear significantly engorged and the overlying mucosa markedly hyperemic

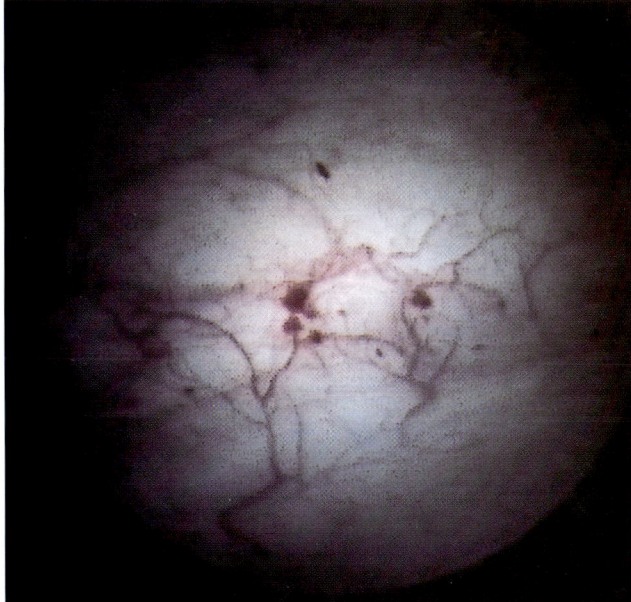

Figure 2.27 Telangiectasias. Distention of the bladder causes intermittent expansion followed by rapid decompression. The elastic properties of the bladder wall allow this expansion to occur without damage in the healthy bladder. However, even in the absence of inflammation, the rapid expansion and decompression of the bladder can cause vascular injury. Vascular fragility is manifested by vessel rupture and pooling of blood around the injured structure. The term telangiectasia describes the appearance of the injured vessel. Telangiectasias are commonly seen during cystoscopy and tend to be focal and located in mobile areas of the bladder such as the bladder dome

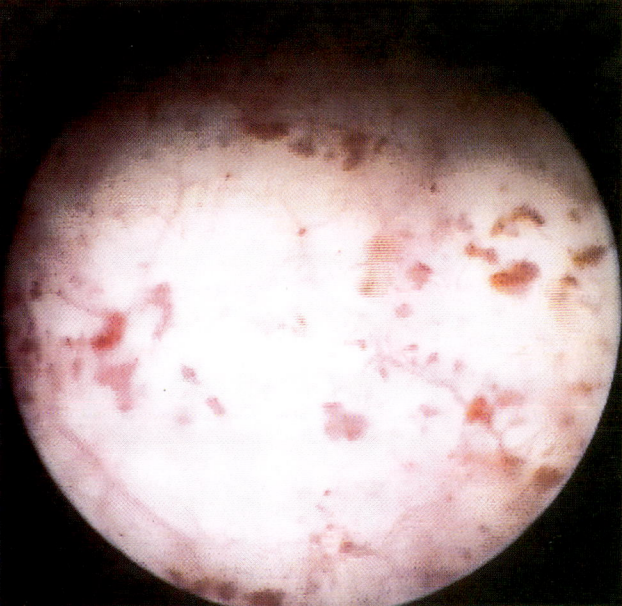

Figure 2.28 In the presence of an acutely inflamed bladder and coexistent with some chronic vesical inflammatory processes, extensive vascular fragility is demonstrable. With distention of the bladder, extensive telangiectasias can be seen which involve all areas of the bladder. This pattern of vascular change is consistent with, but not diagnostic of, interstitial cystitis. These changes can also be seen with other disease entities (such as acute cystitis)

AN ATLAS OF SIGMOIDOSCOPY AND CYSTOSCOPY

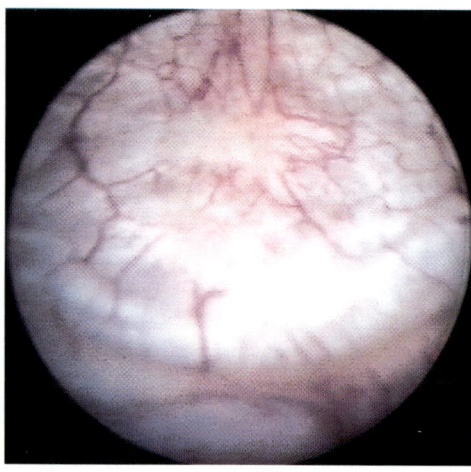

Figure 2.29 Scarring of the urinary bladder wall. The urinary bladder heals rapidly (80% of original tensile strength at 5 days post-injury); however, repeated injury can result in fibrosis of elements within the bladder wall. Bladder biopsy done for histopathological diagnosis of bladder lesions causes mucosal scarring. These scars often have a stellate appearance and may be devoid of submucosal vascularity. Urothelial scars can also be seen after open surgical procedures, such as cystotomy, performed on the bladder

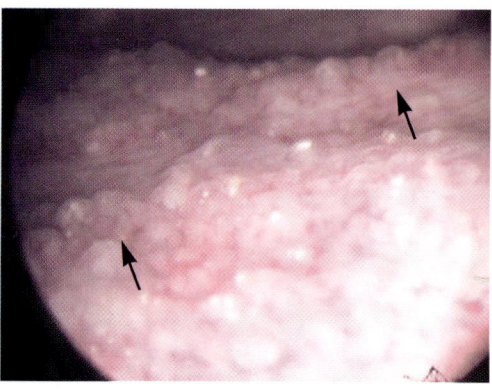

Figure 2.30 Cystitis cystica. Bladder inflammation produces a variety of lesions. The most common inflammatory lesion of the bladder is cystitis cystica, which resembles multiple small cysts within the bladder epithelium. These lesions are usually distributed throughout the bladder, but may be focal and limited to the trigone. The cysts are often multiple and pink to red in color (arrows). Other inflammatory lesions may be located in close proximity to the cystitis cystica inducing cystitis glandularis

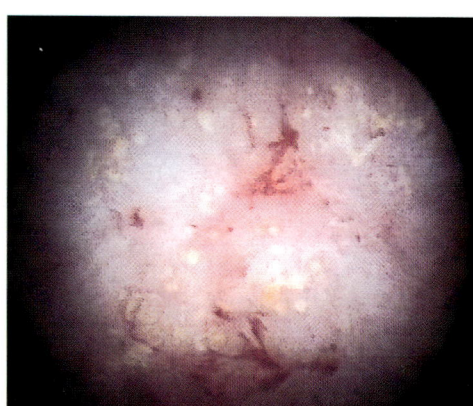

Figure 2.31 Cystitis glandularis. Less common inflammatory lesions of the bladder include cystitis glandularis and follicularis. Both lesions represent chronic inflammatory changes within the bladder and often both are associated with cystitis cystica. These lesions may have a golden hue and also may have a nodular texture. These lesions can be focal or global and are often seen with evidence of mucosal fragility

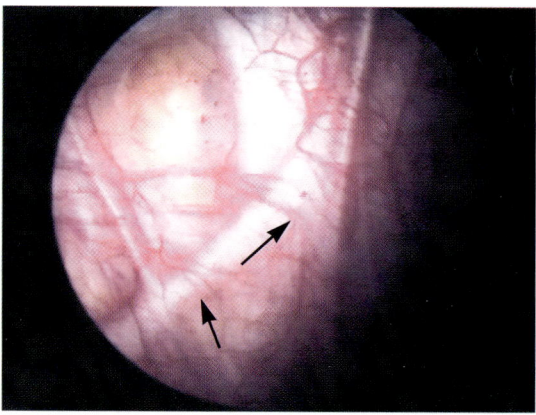

Figure 2.32 Early trabeculation. Bladder muscle responds to certain stimuli by hypertrophy. This hypertrophy increases the thickness of the bladder wall and creates prominent ridges in the smooth muscle. These ridges are called trabeculations and are easily seen during cystoscopy. Trabeculation is graded according to the prominence of the ridges and any abnormality of the intervening bladder wall. Grade I trabeculation (arrows) causes slight elevation of the overlying mucosa

SIGMOIDOSCOPY AND CYSTOSCOPY ILLUSTRATED

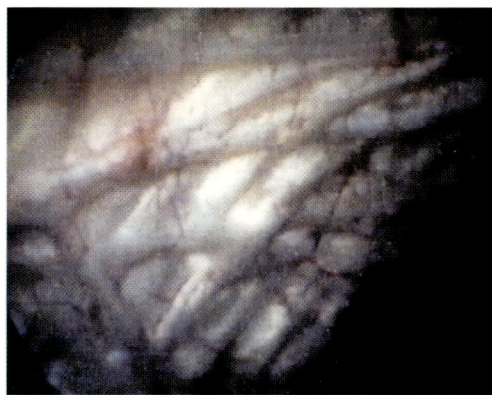

Figure 2.33 Advanced trabeculation. Grade III trabeculation demonstrates even more prominence of the muscular bundles and also a concavity in the intervening mucosa which represents early cellule formation. Cellules are weaknesses in the bladder wall associated with atrophy of the underlying muscle (arrow)

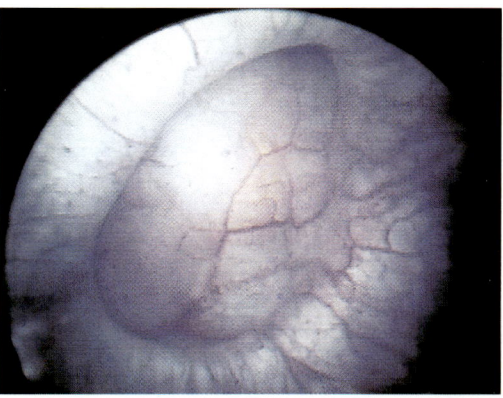

Figure 2.34 Diverticulum. Grade IV trabeculation demonstrates very pronounced muscular bundles with more defined intervening areas of weakness. The intervening areas are called diverticula and represent areas with complete loss of underlying muscle. The bladder wall in these areas is composed of mucosa, lamina propria and underlying adipose tissue only

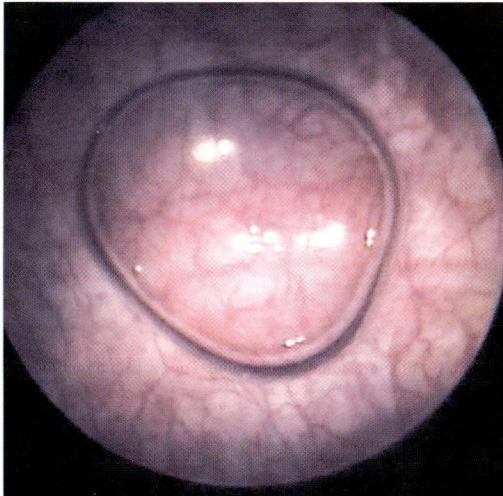

Figure 2.35 Air at bladder dome. Instrumentation of the urinary tract results in the introduction of air into the bladder lumen. Free air in the bladder lumen will rise to the dome of the bladder and create loculated pockets of gas. Other causes of air in the urinary tract include infection with gas-forming organisms and fistula between bowel and urinary tract

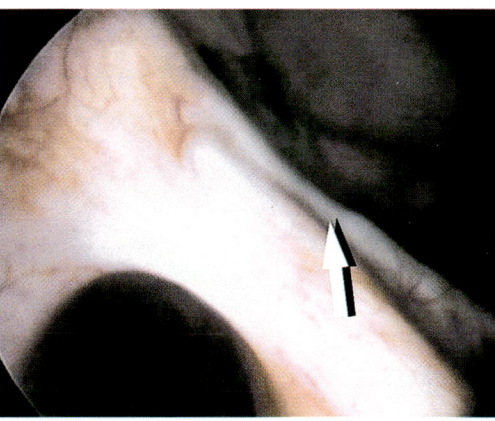

Figure 2.36 Distortion by sutures. One indication for cystoscopy is recurrent incontinence after bladder suspension. Cystoscopy can visualize intraluminal foreign bodies or distortion of the bladder created by errant suture placement. Burch sutures placed in the wall or lumen of the bladder can impinge on the bladder and obstruct ureteral outflow (arrow)

AN ATLAS OF SIGMOIDOSCOPY AND CYSTOSCOPY

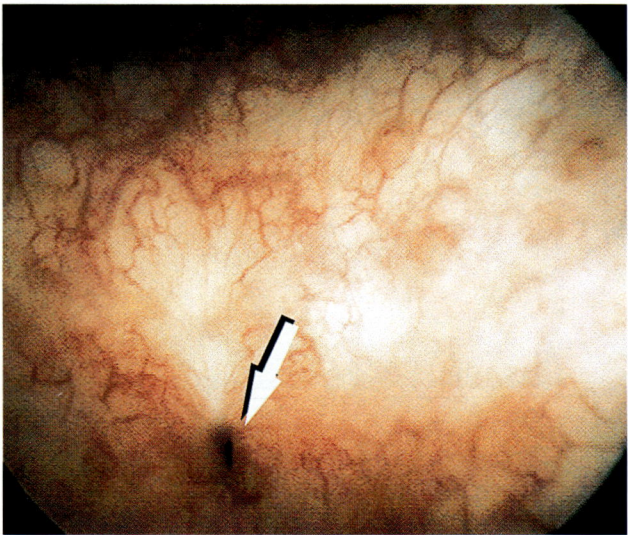

Figure 2.37 Vesicovaginal fistula. Another cause of incontinence after pelvic surgery is fistula between the bladder and vagina. Most commonly, this will be noted on or immediately behind the trigone and will have associated surrounding mucosal hyperemia and edema

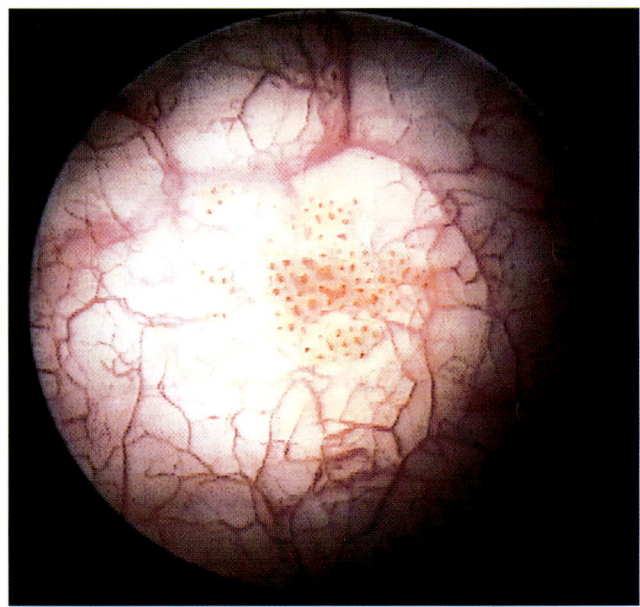

Figure 2.38 Carcinoma *in situ*. Hematuria must be evaluated with cystoscopy to exclude bladder malignancy. Bladder cancer is termed a 'field disease', meaning that the entire urothelium is at risk once carcinoma has been diagnosed. Early hyperplastic changes can be subtle and resemble a clustering of small blood vessels

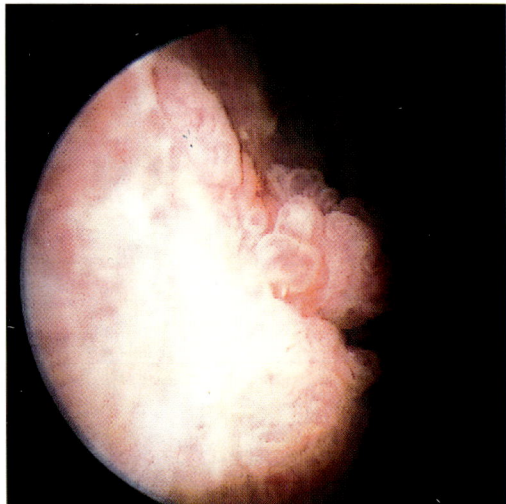

Figure 2.39 Early transitional carcinoma. As a transitional carcinoma develops, it becomes more hyperplastic and also exuberant in its growth pattern. The carcinoma develops a frondular growth pattern and has been likened to a cluster of grapes

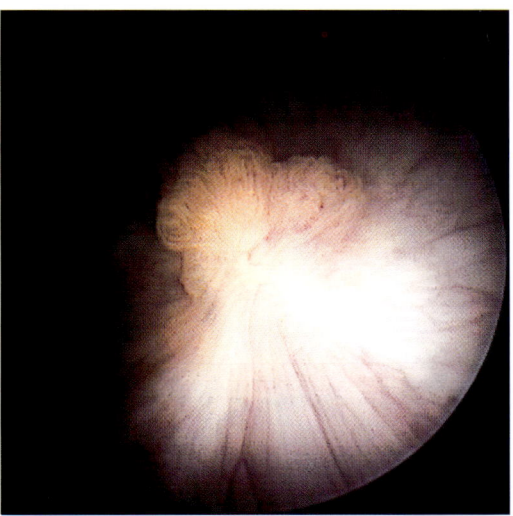

Figure 2.40 Transitional carcinoma. As the tumor continues to grow, more space is occupied. Tumors can occur at any location in the bladder, although often the first tumor presentation is on the trigone and subsequent recurrences occur on the dome or lateral walls. Tumors can also recur in prior areas of resection, as seen here

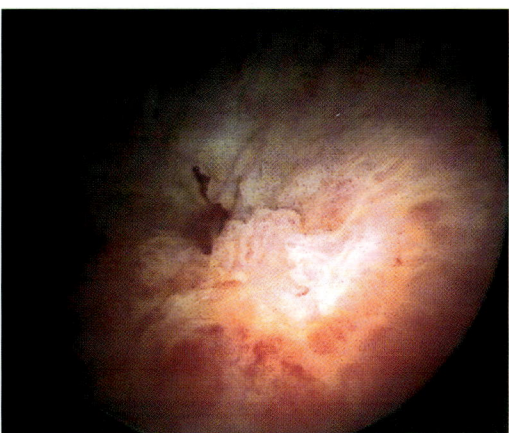

Figure 2.41 Advanced transitional carcinoma. Tumors spread not only outward but also inward. Invasion spreads sequentially through the layers of the bladder, emanating from the mucosa and growing through the lamina propria, muscularis and eventually serosa and perivesical fat. If it is diagnosed before muscular invasion, local resection augmented by intravesical chemotherapy can often cure the disease. If muscular invasion has occurred, as in this case, then cystectomy is indicated for disease control

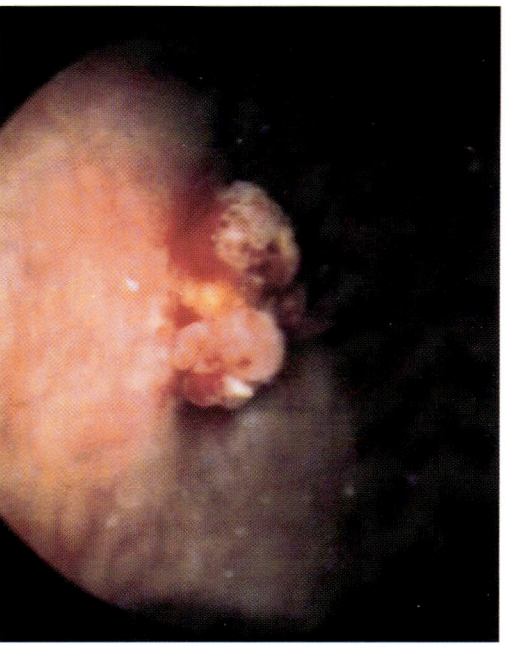

Figure 2.42 Cervix cancer involving the bladder. Rarely a malignancy of the uterine cervix grows anteriorly and involves the bladder base, often close to the trigone. Ureteral obstruction is frequent in this situation. It is often impossible to pass a stent, owing to significant distortion of the lower ureter and the ureteral orifice

Index

A
adenocarcinoma
  rectum 85
  sigmoid 88
adenoma
  and colorectal cancer 15
  proximal sigmoid 16
adhesions, sigmoidoscopy examination 46
anal canal, anal papilla and dentate line 83
anatomy
  for cystoscopy 65–6, **90**
  for sigmoidoscopy 41–2, 79, **80, 83**
antibiotic prophylaxis 34, 62, 63
anticoagulation 34–5, 62
anus and rectum
  anatomy 41–2, 79, **83**
  perianal fissure **85**
anxious patient 33–4, 61
aphthous ulcer, Crohn's disease of colon **84**
arteriovenous malformation **88**

B
bacterial endocarditis, prevention 34, 62
barium enema
  diverticular disease 21
  indications for sigmoidoscopy 18, 22
biopsy instruments
  cystoscopy 58, 69, **89**
  sigmoidoscopy 28, 31
bladder
  acute inflammation **95**
  air at bladder dome **97**
  anatomy 65, **90**
  biopsy 58, 69, **89**
  carcinoma *in situ* **98**
  cervical cancer involving **99**
  cystoscopy examination 65–72, **88–99**
  distortion by sutures **97**
  diverticulum **97**
  evaluation 68–9
  fistula 54, 70
  mural vessels **95**
  normal mucosa **95**
  squamous metaplasia **94**
  telangiectasias **95**
  trabeculation
    advanced **97**
    early **96**
  transitional carcinoma **98**
    advanced **99**
    early **98**
  trigone **93**
  tumours
    follow-up 54
    hematuria 53–4
  wall scarring **96**
    sessile polyps **85**
bladder neck **93**
bladder stones 54
bowel preparation for sigmoidoscopy 35–6
  colonoscopy 36
  constipation 36
  elderly patients 36
  inflammatory bowel disease 36
  medications 36
  pregnant patients 36
  special procedures 36

C
camera, sigmoidoscopy 28–9
cardiac problems
  antibiotic prophylaxis 34, 63
  complicating sigmoidoscopy 47
cervical cancer
  colonoscopy 22
  involving bladder **99**
charge coupled device chip (CCD) 25, 28
Clostridium infection/contamination 29
colitis
  chronic idiopathic ulcerative (pan)colitis 18–19, 21, **86, 87**
  indications for sigmoidoscopy 21–2
  infectious 21
  ischemic colitis with ulceration 27, **87**
  proctitis 21–2

INDEX

pseudomembranous colitis 21, **86**
colon
   anatomy 41–2
   arteriovenous malformation **88**
   descending colon 42, **80**
   normal mucosa **86**
colonic diverticulum **88**
colonic pedunculate polyps **88**
colonoscopy
   bowel preparation 36
   cervical cancer 22
   oximetry 35
colorectal cancer
   HNPCC, Lynch I 18
   preclinical phase 15
   prevalence 16–17
   rectum **85**
   recurrence 18
   screening 15–19
      age at commencement 17–18
      family history of colon cancer 18
      fecal occult blood testing (FOBT) 16, 18–19
      long preclinical phase 15
      screening intervals 18
      treatment results and validity 15, 16
   sigmoid **88**
consent forms 36–8, 63, 72
Crohn's disease of colon **86**
   aphthous ulcer **84**
   perianal fissure **85**
   rectal mucosal tags **84**
   rectovaginal fistula **85**
cystitis 53
   cystitis cystica **96**
   cystitis glandularis **96**
cystoscopy
   report *71*
   training 72
cystoscopy equipment 57–9, **88–90**
   biopsy instruments 58, 69, **89**
   cystourethroscopes, flexible/rigid 57–8, **89**
   history 57
   lens systems 58, 67
   light sources 58–9
   maintenance and sterilization 59
   monitors 58
   sheath 67
cystoscopy examination 65–72, **88–99**
   complications 72
      bleeding 70
   distension media 66–7
   documentation 72
   fistula evaluation 54, 70
   intraoperative cystoscopy 69–70
   male patients 72
   preoperative cystoscopy 54
   preparation of patient 61–3, 66
   procedure and quitting 67–72
   suprapubic approach 55, **91**
cystoscopy indications 53–63
   atypical urinary tract infection 53
   bladder stones 54
   contraindications 55
   fistulas 54, 70
   follow-up after bladder tumours 54
   hematuria 53–4
   incontinence 53
   interstitial cystitis 53
   intraoperative cystoscopy 54–5
   preoperative cystoscopy 54
   staging of pelvic malignancies 54

D
diabetic patients 35
diarrhea
   infectious colitis 21
   pseudomembranous colitis 21, **86**
disinfection/decontamination 29–31
distension media 66–7
diverticula
   colonic **88**
   cystoscopy indications 53
   urethral **93**
diverticular disease 21
diverticulosis
   and diverticulitis 21, **87**
   sigmoidoscopy examination 46
   with small lesions **87**
   thickened mucosal folds **87**
documentation
   cystoscopy examination 72
   sigmoidoscopy examination 48, *49*

E
elderly patients 34, 61
enema, bowel preparation for sigmoidoscopy 35–6
examination room 59

F
familial polyposis 19
fecal occult blood testing (FOBT) 16, 18–19
female urethral meatus, normal **91**
fiberoptics
   cystoscopy 58–9
   sigmoidoscopy 27–8
fistula
   anorectal, evaluation 22
   rectovaginal fistula **85**
   vesicovaginal fistula 70, **98**

G
glutaraldehyde, disinfection/decontamination of equipment 30–31
gut lavage, bowel preparation for sigmoidoscopy 35–6
gynecological malignancies 22, 54

H
hematuria, indications for cystoscopy 53–4
hemorrhoids
   small internal **84**
   thrombosed external **84**
hereditary non-polypotic colon cancer (HNPCC) 18

history of cystoscopy 57
HIV infection 22
Hunner's ulcers 53

I
incontinence 53
indications
   for cystoscopy *see* cystoscopy indications
   for sigmoidoscopy *see* sigmoidoscopy indications
infection/contamination of equipment 29–31
inflammatory bowel disease
   bowel preparation 36
   and colorectal cancer 18, 19
instructions to patient
   cystoscopy 62, 63
      post-examination instructions 72
   sigmoidoscopy 36, *37*
      post-examination instructions 48
instruments *see* cystoscopy equipment;
      sigmoidoscopy equipment
insufflation 26
interstitial cystitis 53
intraoperative cystoscopy 69–70
   indications 54–5
iron medication, preparation for sigmoidoscopy 36
ischemic colitis with ulceration 27, 87

L
light sources 28
   cystoscopy 58–9
   sigmoidoscopy 28
lumen, location in sigmoidoscopy 44–5, 81, 82

M
melanosis coli **86**
monitors
   cystoscopy 58
   sigmoidoscopy 35
Mycobacterium infection/contamination 29
myocardial infarction 34

P
patient history
   cystoscopy 61
   sigmoidoscopy 33–5
patient preparation for cystoscopy 61–3
   antibiotic prophylaxis 62, 63
   anticoagulation 62
   diabetic patients 62
   elderly patients 61
   instructions 62, 63, 73
   localk anesthesia 66
   pregnant patients 61
patient preparation for sigmoidoscopy 33–8
   bowel preparation 35–6
   consent 36–8, *38*
   instructions 36, *37*, 48
   laboratory evaluation 35
   physical examination 35
pelvic malignancies 22
   staging 54
pelvic mass 46

pelvic surgery, adhesions 46
perforation
   contraindications to sigmoidoscopy 22–3
   sigmoidoscopy examination 47–8
perianal fissure **85**
polyps
   colonic pedunculate **88**
   familial polyposis 19
   inflammatory or pseudopolyps **87**
   rectal **83**, **85**
pregnancy 34
   bowel preparation for sigmoidoscopy 36
pseudomembranous colitis 21, **86**
Pseudomonas infection/contamination 29

R
radiation proctitis **86**
rectal adenocarcinoma **85**
rectal anatomy 79
rectal bleeding 19–21
   after pelvic radiation 21
   causes *20*
   post sigmoidoscopy 48
   severe 20–1
   subclinical 20
rectal mucosal tags **84**
rectal polyps
   dentate line **83**
   hyperplastic polyps **85**
   pedunculate **85**
rectal ulcer syndrome **84**
rectal veins **83**
rectovaginal fistula **85**
resuscitation kit 33

S
Salmonella infection/contamination 29
screening for colorectal cancer 15–19
sedation 33–4
sigmoid
   adenocarcinoma **88**
   proximal, and adenoma 16
sigmoid anatomy 42
   inflammatory or pseudopolyps **87**
   pedunculate polyps **88**
   variations in form and position **80**
sigmoidoscopy equipment 25–32
   accessories 28–9
   biopsy forceps 28, 31
   buying 29
   cleaning/maintenance/storage 29–31
   endoscopic shaft tip 77
   flexible/rigid 25–6
   light sources 28
   set–up 31–2, **78**
   types
      fiberoptics 27–8
      length/diameter 26–8
      video sigmoidoscope 28, 77
sigmoidoscopy examination 41–50
   anatomy of lower GI tract 41–2, 79
   bowel preparation 35–6

INDEX

complications 46–8
  bleeding 48
  cardiac problems 47
  medical 47
  perforation 47–8
  vagal reactions 47
documentation 48, *49*
positioning the patient 42
problem areas 46
  adhesions 46
  diverticulosis 46
  pelvic mass 46
  severe inflammation 46
procedure and quitting 45–6
screening
  cost of screening 16–17
  ease of administration 16
  length of scope used 17
  patient selection 17, 18, 19
  physicians involved 18–19
  serious disease 15
  toleration of screening 16
training 48, 50
sigmoidoscopy indications 15–23
  abnormal barium enema 22
  colitis 21–2, 27, **86**, **87**
  contraindications 22–3
  diverticular disease 21
  fistula 22
  pretreatment, gynecological malignancies 22
  proctitis 21–2, **84**
  rectal bleeding 19–21
  symptomatic patients without evidence of bleeding 22
sigmoidoscopy technique
  endoscope handling 42–4
    air insufflation 44
    basic movements 42–3, **81**
    control of movement around a bend **82**
    finding lumen 43, 44–5, **81**, **82**
    irrigation/suction 44
    one vs two hands 42, 80
    orientation 44–5, **81**, **82**
    N–loops 44–5, **82**, **83**
    twisting/torque 43
      passing sigmoid **81**
    impacted scope 81
solitary rectal ulcer syndrome **84**
squamous carcinoma, urethra **91**

T
telangiectasias, bladder **95**
training
  cystoscopy 72
  sigmoidoscopy 48, 50

U
ulcer, Crohn's disease of colon **84**
ulcerative colitis 18–19, 21, **86**, **87**
ulcerative proctitis **84**
ureteral anatomy 66
  evaluation 68–9
ureteral lesions **94**
ureteral orifice, abnormal/normal **94**
urethra
  anatomy 65–6
    evaluation 68
  diverticulum **93**
  dynamic urethroscopy 688
  inflammation **93**
  inflammatory polyps **92**
  mucosal prolapse **91**
  normal mucosa **92**
  squamous carcinoma **91**
urethral meatus, normal **91**
urinary tract infection 53

V
vagal reactions, sigmoidoscopy examination 47
vesicovaginal fistula **98**
video sigmoidoscope 28–9, 77